1 MONTH OF
FREE
READING

at

www.ForgottenBooks.com

By purchasing this book you are eligible for one month membership to ForgottenBooks.com, giving you unlimited access to our entire collection of over 1,000,000 titles via our web site and mobile apps.

To claim your free month visit:

www.forgottenbooks.com/free32716

ISBN 978-1-5283-5140-9
PIBN 10032716

CHEMICAL

PROBLEMS AND REACTIONS,

TO ACCOMPANY

STÖCKHARDT'S ELEMENTS OF CHEMISTRY.

BY

JOSIAH P. COOKE, Jr.

CAMBRIDGE:

PUBLISHED BY JOHN BARTLETT,

Bookseller to the University.

1857.

CAMBRIDGE:

ELECTROTYPED AND PRINTED BY METCALF AND COMPANY.

PREFACE.

THIS book has been prepared solely for the use of the undergraduates of Harvard College. It contains a collection of chemical problems and reactions, with references to the sections of Stöckhardt's Elements of Chemistry, and also a few chapters on the chemical nomenclature and the use of chemical symbols, subjects which are not sufficiently developed in that text-book for the purposes of college instruction. In writing chemical symbols the author has adopted a uniform system throughout the volume, which, as he hopes, will be found to be at once expressive and clear. The problems and reactions cover the Inorganic portion of Stöckhardt's Elements ; the problems have only been extended to the section on the Heavy Metals. Beyond this, the reactions alone have been given, as it was supposed that, before reaching this section, the student will easily be able to propose problems for himself. In solving many of the problems it will be found convenient to use logarithmic tables of four places, which, with several other tables, will be found at the end of the volume. The student is advised to remove the tables of logarithms, and paste them for use on a card. The difficulty of insuring complete accuracy in the printing of chemical formulæ can be known only

press. Several errors have been already discovered, and corrected, but others unquestionably exist.

CAMBRIDGE, *May* 15*th*, 1857.

NOMENCLATURE OF CHEMISTRY.

Origin of the Nomenclature. — Previous to the year 1787 the names given by chemists or alchemists to substances were not conformed to any general rules. Many of these old names, such as *Oil of Vitriol, Calomel, Corrosive Sublimate, Red-Precipitate, Saltpetre, Liver of Sulphur, Cream of Tartar, Glauber's and Epsom Salts,* are still retained in common use. As chemical science advanced, and the number of known substances increased, it became important to adopt a scientific nomenclature. The admirable system now in use is due almost entirely to Lavoisier, who reported to the French Academy on the subject, in the name of a committee, in 1787. This system, now known as the Lavoisierian nomenclature, was generally adopted by scientific men soon after its publication, and has not been materially modified since. In it the name of a substance is made to indicate the composition.

Names of the Elements. — The names of the elements are the only ones which are now independent of any rule. Those which were known before the adoption of the nomenclature,

1

such as Sulphur, Phosphorus, Iron, Lead, retain their old names. Several of the more recently discovered elements have been named in allusion to some prominent property or some circumstance connected with their history; as, *Oxygen* from ὀξύς, γεννάω (acid-generator); Hydrogen, from ὕδωρ, γεννάω (water-generator); Chlorine, from χλωρός (green); Iodine, from ἰωδής (violet); Bromine, from βρῶμος (fetid odor), &c. The names of the newly discovered metals have a common termination, *um*, as Platinum, Potassium, Sodium; and the names of a class of the metalloids terminate in *ine*, as Chlorine, Bromine, &c.; but except in these respects the names of the elements are entirely arbitrary.

Classification of Compounds. — There are three orders of chemical compounds : — 1st, *Binary Compounds*, consisting of two elements, or of the representatives of two elements; 2d, *Ternary Compounds*, consisting of three elements, or of their representative; and 3d, *Quaternary Compounds*, consisting of four elements, or their representatives. There are some chemical compounds containing more than four elements; but in most cases two or more of these elements are representatives, i. e. occupy the place, of only one, as will be explained farther on. Binary compounds are subdivided into two classes, *Electro-Positive Binaries*, or *Bases*, and *Electro-Negative Binaries*, or *Acids*. Each of these classes is distinguished by a peculiar set of properties, or at least this is the case with the prominent members of either class; but the two classes merge so gradually into each other, that it is impossible to draw a line of demarcation between them; and there is a large class of intermediate compounds, which either partake of the properties of both, or are entirely indifferent. Indeed, the binary compounds may best be regarded as forming a continuous series of substances, varying in their properties from those of strong acids on the one hand to those of strong bases on the other, and with every possible grade of qualities between the two extremes. In this series each binary may be considered as an electro-positive compound or base towards all those which precede it, and as an electro-negative compound or acid towards all those which follow it. Ternary compounds are generally, at least in Inorganic Chemistry, composed of two binaries, i. e. of an acid and a base, and are then called Salts. The quaternary compounds are generally composed of two salts, and are called Double Salts.

Names of Binaries. — The most important binaries, as well as those which have been the best studied, are the compounds of oxygen with the other elements. To these the generic term Oxide has been applied. The electro-positive binaries are called simply Oxides of the elements of which they consist. Thus we have

Oxide of Hydrogen, consisting of oxygen and hydrogen.
Oxide of Potassium,   "   "   "   potassium.
Oxide of Sodium,   sodium.

When the name of the metal ends in *um*, the name of the compound with oxygen is frequently formed by changing this termination into *a*, with such other modifications of the terminal letters as euphony may require. Thus we use, instead of

Oxide of Sodium,   Soda.
Oxide of Potassium,   Potassa.

Oxide of Calcium,	Calcia (or Lime).*
Oxide of Barium,	Baryta.
Oxide of Strontium,	Strontia.
Oxide of Magnesium,	Magnesia.
Oxide of Aluminum,	Alumina.

The two names are in all cases synonymous. Generally oxygen combines with an element in more than one proportion; then, in order to distinguish between the different oxides of the same element, we use various Latin and Greek prefixes, such as *sub, proto, sesqui, deuto, hyper.* This is well illustrated by the names of the different oxides of mercury and manganese, which are as follows.

	Composition.	
Names.	Mercury.	Oxygen.
*Sub*oxide of Mercury	100	4
*Prot*oxide of Mercury	100	8

	Manganese.	Oxygen.
*Prot*oxide of Manganese	27.6	8
*Sesqui*oxide of Manganese	27.6	12
*Hyper*oxide of Manganese	27.6	16

The electro-negative binary compounds of oxygen (the acids) are named on a different principle. These are called different kinds of acids. If the element forms but one acid with oxygen, the name is formed by adding to the name of the acid the termination *ic*, with such changes of the final letters as euphony may require. Thus, carbon and oxygen form Carbon*ic*

---

* The common name *Lime* is much more frequently used than either of its scientific synonymes, Oxide of Calcium, or Calcia. Indeed, the last has never been in general use.

Acid. When the element forms two acids, by combining with different amounts of oxygen, the termination *ic* is reserved for that containing the most oxygen, while the termination *ous* is given to the other. We have, for example,

	Arsenic.	Oxygen.
Arseni*ous* Acid	75	24
Arsen*ic* Acid	75	40

	Phosphorus.	Oxygen.
Phosphor*ous* Acid	32	24
Phosphor*ic* Acid	32	40

If oxygen combines with an element in more than two proportions, to form acids, the names are formed with the Greek prefix *hypo*, indicating a less, or the Latin prefix *per*, indicating a greater, amount of oxygen than that contained in the acids to whose names they are prefixed. The acid compounds of sulphur and oxygen are

	Sulphur.	Oxygen.
*Hypo*sulphur*ous* Acid	16	8
Sulphur*ous* Acid	16	16
*Hypo*sulphur*ic* Acid	16	20
Sulphur*ic* Acid	16	24

The acid compounds of chlorine and oxygen are

	Chlorine.	Oxygen.
*Hypo*chlor*ous* Acid	35.5	8
Chlor*ous* Acid	35.5	24
*Hypo*chlor*ic* Acid	35.5	32
Chlor*ic* Acid	35.5	40
*Per*chlor*ic* Acid	35.5	56

Very frequently the higher degrees of oxidation of an ele-

1 *

ment are acids, when the lower degrees are bases, or indifferent
compounds. This is the case with the oxides of manganese.
Besides the three already mentioned, there are also

	Manganese.	Oxygen.
*Mang*an*ic* Acid	27.6	24
*Per*mangan*ic* Acid	27.6	28

The different oxides of nitrogen are, in like manner,

	Nitrogen.	Oxygen.
*Prot*oxide of Nitrogen	14	8
*Deut*oxide of Nitrogen	14	16
Nitr*ous* Acid	14	24
*Hypo*nitr*ous* Acid	14	32
Nitr*ic* Acid	14	40

It will be noticed that here, as in other places, the term
oxides is used generically for all the compounds of an element
with oxygen, whether acids or bases, and the terms *protoxide*,
&c. to designate the specific compounds. It is not unfrequently the case that the same oxide acts as an acid under
some circumstances, and as a base under others, and accord-
ingly is known under two different names. Water, when a
base, is called *Oxide of Hydrogen*, but when an acid, it is
named *Hydric Acid*. The oxide of aluminum, when a base,
is called *Sesquioxide of Aluminum*, but when an acid, *Alu-
minic Acid*.

Next to the oxides, the compounds of sulphur with the
elements are the most important binaries. These are called
sulphides, and, as a general rule, for every oxide there is a
corresponding sulphide, which is named after the analogy of
the name of the oxide. Thus we have two sulphides of mercury : —

	Mercury.	Sulphur.
*Sub*sulphide of Mercury	100	8
*Proto*sulphide of Mercury.	100	16

There are three sulphides of iron :—

	Iron.	Oxygen.
*Proto*sulphide of Iron	28	8
*Sesqui*sulphide of Iron	26	12
*Per*sulphide of Iron	28	16

The last corresponds to no oxide. The sulphides, like the oxides, may be divided into basic or indifferent sulphides, and into acid sulphides. The sulphur bases are named as above. The sulphur acids have been named by prefixing to the name of the oxygen acid the letters *sulpho*. Thus we have, cor‧responding to

Arsenious Acid,	*Sulpho*arsenious Acid.
Arsenic Acid,	*Sulpho*arsenic Acid.
Carbonic Acid,	*Sulpho*carbonic Acid.
Hydric Acid,	*Sulpho*hydric Acid.

The names of the binary compounds of the other elements are named after the same analogy as the oxides and sulphides. Thus with the other elements

Oxygen	forms	Oxides.
Fluorine	"	Fluorides.
Chlorine		Chlorides.
Bromine		Bromides.
Iodine		Iodides.
Sulphur		Sulphides.
Selenium		Selenides.

Tellurium	forms	Tellurides.
Nitrogen	"	Nitrides.
Phosphorus	"	Phosphides.
Arsenic		Arsenides, &c.

We distinguish the different fluorides, bromides, &c. by prefixes, as we do the oxides. Each of these classes of compounds may be subdivided, like the oxides or sulphides; but only in a few cases has the nomenclature of the oxygen and sulphur acids been extended to them. Almost the only instances are the names of the compounds of the first few elements of the last list with hydrogen, which are sometimes called Fluohydric, Chlorohydric, Bromohydric Acids, &c.; but even here the synonymous names Hydrofluoric, Hydrochloric, Hydrobromic Acids, &c. are more generally used.

Names of Ternaries. — The name of a ternary compound, or salt, is very simply formed from the names of the acid and base of which it consists. The termination of the name of the acid is changed, if *ic*, into *ate;* if *ous*, into *ite ;* and the name of the base added. Thus, sulphur*ic* acid and oxide of sodium form Sulph*ate* of the Oxide of Sodium (or of Soda); sulphur*ous* acid, Sulph*ite* of the Oxide of Sodium. In like manner, hyposulphur*ic* acid forms Hyposulph*ate*, and hyposulphur*ous* acid, Hyposulph*ite* of the Oxide of Sodium. So, also, Chlorite of Oxide of Potassium, and Chlorate of Oxide of Potassium; Selenite of Oxide of Barium, and Selenate of Oxide of Barium; Nitrite of Oxide of Calcium, and Nitrate of Oxide of Calcium; Arsenite of Oxide of Lead, and Arsenate of Oxide of Lead. When the base is an oxide of one of the heavy metals, the name of the salt is frequently abbreviated by leaving out the word *oxide;* as, Sulphate of Lead, for Sulphate of the Oxide of Lead; and Nitrate of Copper, for Nitrate of the Oxide of Copper. When the base is an oxide of one of the light metals, by substituting for Oxide of Sodium, Oxide of Potassium, &c., the single words Soda, Potassa, &c., as already explained on page 3. Instead of Sulphate of Oxide of Barium, chemists more frequently write the shorter Sulphate of Baryta; instead of Nitrate of Oxide of Calcium, Nitrate of Lime. When the same acid combines with two or more different oxides of the same element to form salts, these are distinguished by introducing the specific name of the oxide into the name of the salt; as, Sulphate of *Prot*oxide of Iron, and Sulphate of *Sesqui*oxide of Iron; Nitrate of *Sub*oxide of Mercury, and Nitrate of *Prot*oxide of Mercury. It is frequently the case that an acid combines with the same base in two or more different proportions, forming two or more different salts. In order to distinguish these salts, the one is called the *neutral* salt; that containing more base than the neutral salt, a *basic;* and that containing less, an *acid.* Thus we have

> Basic Phosphate of Lime.
> Neutral Phosphate of Lime.
> Acid Phosphate of Lime.

So, also,

> Basic Chromate of Lead.
> Neutral Chromate of Lead.

When there are several basic salts, they are distinguished by Latin prefixes. Thus there are five acetates of lead : —

	Oxide of Lead.	Acetic Acid.
Neutral Acetate of Lead	112	51
Bibasic Acetate of Lead	224	51
Sesquibasic Acetate of Lead	280	51
Tribasic Acetate of Lead	336	51
Sexbasic Acetate of Lead	672	51

The acid salts are distinguished by placing the Latin prefixes directly before the name of the neutral salts; as,

	Soda.	Boracic Acid.
Borate of Soda	31	35
Biborate of Soda	31	70

	Potassa.	Chromic Acid.
Neutral Chromate of Potassa	47	51
Bichromate of Potassa	47	102
Trichromate of Potassa	47	153

Most of the oxygen acids are commonly used only when in combination with water. These compounds are true salts, in which the water plays the part of a .base. The common Sulphuric Acid is a Sulphate of Water, and the common Nitric Acid, Nitrate of Water. Properly speaking, the terms Sulphuric and Nitric Acid ought only to be applied to the anhydrous compounds; but custom authorizes us to extend these names to the hydrates.

The names of the ternary sulphur compounds, the sulphur salts, are formed in the same way as those of the oxygen salts; thus, the compound of sulphocarbonic acid and sulphide of sodium is called the Sulphocarbonate of the Sulphide of Sodium; the compound of sulphoarsenic acid and sulphide of potassium, the Sulphoarsenate of the Sulphide of Potassium; and the compound of sulphoarsenious acid and the same base, Sulphoarsenite of the Sulphide of Potassium.

The names of the ternary fluorine, chlorine, bromine, and iodine compounds are formed after a different analogy. These are generally regarded as double salts, and are called the double fluorides, chlorides, bromides, or iodides of the two metals which they contain. Hence the names

Double Chloride of Aluminum and Sodium.

Double Chloride of Platinum and Ammonium.

Names of Quaternaries. — The double salts formed from two salts each containing the same acid but different bases, are called double salts of the two bases. Thus, the compound of sulphate of alumina and sulphate of potassa is called the Double Sulphate of Alumina and Potassa; the compound of sulphate of soda and sulphate of zinc, the Double Sulphate of Soda and Zinc. So, also, the Double Sulphate of Potassa and Magnesia, and the Double Sulphate of Potassa and Water. The last is more frequently called Bisulphate of Potassa, as if it were composed of one equivalent of base and two equivalents of acid, instead of being, as it probably is, a compound of two salts. Similar incongruities in the nomenclature, arising frequently from different theories in regard to the constitution of substance, are not uncommon among the higher orders of compounds.

The Lavoisierian nomenclature has not been found sufficiently expansive to meet the requirements of modern science, not only on account of the complex-character of many of the newly discovered compounds, but more especially because it involves the ideas of a peculiar theory, which, although once almost universally received, is not now so generally admitted. The inadequacy of the old nomenclature to afford names for the great variety of chemical compounds, or to describe the infinite changes to which they are liable, has given rise to a peculiar chemical language, analogous to the language of mathematics, called Chemical Symbols. To explain the use of this language will be the object of the next chapter.

CHEMICAL SYMBOLS.

SINCE all matter is composed of one or more of sixty-two different substances, never as yet decomposed, and hence called Elements, it is evident that, if we adopt an arbitrary symbol for each of these elements, we shall be able, by combining them together, to express all possible varieties of combination. Moreover, since the elements always combine in certain fixed and definite proportions by weight, it is equally evident that, if we assign to each of these symbols a certain weight, we shall be able to indicate the relative quantities of the different elements which enter into any compound.

Symbols of Elements. — It has been agreed by chemists of different nations to use, as symbols of the elements, the first letters of their Latin names. When two or more names commence with the same letter, a second letter is added for distinction. The first letter is printed or written in capitals, and the second, when used, in small letters, immediately following the first. A list of the Elements, with their Symbols, is given on the following page.

2

CHEMICAL SYMBOLS AND EQUIVALENTS.

Aluminum	Al = 13.7	Nickel	Ni = 29.6
Antimony (Stibium)	Sb = 129	Niobium	Nb
Arsenic	As = 75	*Nitrogen*	N = 14
Barium	Ba = 68.5	Norium	No
Bismuth	Bi = 213	Osmium	Os = 99.6
Boron	B = 10.9	*Oxygen*	O = 8
Bromine	Br = 80	Palladium	Pd = 53.3
Cadmium	Cd = 56	Pelopium	Pe
Calcium	Ca = 20	*Phosphorus*	P = 32
Carbon	C = 6	Platinum	Pt = 98.7
Cerium	Ce = 47	*Potassium* (Kalium)	K = 39.2
Chlorine	Cl = 35.5	Rhodium	R = 52.2
Chromium	Cr = 26.7	Ruthenium	Ru = 52.2
Cobalt	Co = 29.5	Selenium	Se = 39.5
Copper (Cuprum)	Cu = 31.7	*Silicon*	Si = 21.3
Didymium	D	*Silver* (Argentum)	Ag = 108.1
Erbium	E	*Sodium* (Natrium)	Na = 23
Fluorine	Fl = 18.9	Strontium	Sr = 43.8
Glucinum	G = 4.7	*Sulphur*	S = 16
Gold (Aurum)	Au = 197	Tantalum	Ta = 184
Hydrogen	H = 1	Tellurium	Te = 64.2
Iodine	I = 127.1	Terbium	Tb
Iridium	Ir = 99	Thorium	Th = 59.6
Iron (Ferrum)	Fe = 28	*Tin* (Stannum)	Sn = 59
Lanthanium	La	Titanium	Ti = 25
Lead (Plumbum)	Pb = 103.7	Tungsten (Wolfram)	W = 95
Lithium	Li = 6.5	Uranium	U = 60
Magnesium	Mg = 12.2	Vanadium	V = 68.6
Manganese	Mn = 27.6	Yttrium	Y
Mercury (Hydrargyrum)	Hg = 100	*Zinc*	Zn = 32.6
Molybdenum	Mo = 46	Zirconium	Zr = 22.4

The student will do well to notice in the foregoing list the symbols of those elements whose Latin names commence with letters differing from the initial letters of the English names, since they are not so easily remembered as the others.

Chemical Equivalents. — The chemical symbols not only stand for the names of the elements, but also for a fixed proportional weight of each. These weights are given in the above table, opposite to the symbols. They have only relative values; if one is in pounds, all the rest are in pounds; and if one is in ounces, all the rest are in ounces. We may leave the standard indefinite, and express the weight in parts; then Al stands for 13.7 parts of aluminum; Sb stands for 129 parts of antimony, &c. The weight of an element indicated by its symbol is called *one equivalent*, and it is a law of chemistry that elements always combine by equivalents; that is, one equivalent of one combines with one equivalent of another, or else several equivalents of one combine with one or with several equivalents of another.

As stands for 75 parts of Arsenic, or one equivalent.
Bu   "   68.5   "   Barium,   "   "
H   1   "   Hydrogen,   "   "
O   "   8   "   Oxygen,   "   "

In order to express two, three, or more equivalents of an element, we place a figure just below the symbol at its right hand; thus,

O_2 stands for 16 parts of Oxygen, or two equivalents.
O_3 "   24   "   "   three equivalents.

These figures merely multiply the symbols beneath which

they stand, and must not be confounded with algebraic powers, which are sometimes written in a similar way. We sometimes place the figure, though larger, before the symbol; 2 O means exactly the same thing as O_2.

Symbols of Compounds. — In order to form the symbol of a compound, we write the symbols of the elements of which it consists one after the other, indicating by means of figures the number of equivalents of each which have entered into combination. Thus, H O is the symbol of water, a compound consisting of one equivalent or one part of hydrogen, and of one equivalent or eight parts of oxygen; $S O_3$ is the symbol of sulphuric acid, a compound consisting of one equivalent or sixteen parts of sulphur, and of three equivalents or twenty-four parts of oxygen; $C_{12} H_{11} O_{11}$ is the symbol of common sugar, a compound consisting of twelve equivalents of carbon, eleven equivalents of hydrogen, and eleven equivalents of oxygen.

Binary Compounds. — The symbols of binary compounds are formed by writing the symbols of the two elements together, *taking care to place the symbol of the metal, or of the most electro-positive element, first.* The binary symbol thus obtained represents always one equivalent of the compound. The weight of this equivalent is evidently the sum of the weights of the equivalents of the elements entering into the compound.

$14 + 40 = 54.$

$N O_5$ stands for one equiv. or 54 parts of Nitric Acid.

$16 + 24 = 40.$

$S O_3$ "   40   "   Sulphuric Acid.

$6 + 16 = 22.$

C O_2 stands for one equiv. or 22 parts of Carbonic Acid.

$1 + 8 = 9.$

H O	"	9	"	Water.

$28 + 8 = 36.$

Fe O	"	36	"	Oxide of Iron.

$20 + 8 = 28.$

Ca O	"	28	"	Lime.

$23 + 8 = 31.$

Na O	"	31	"	Oxide of Sodium.

$6 + 32 = 38.$

C S_2	"	38	"	Sulphocarbonic Acid.

$75 + 48 = 123.$

As S_3	"	123	"	Sulphoarsenious Acid.

$28 + 16 = 44.$

Fe S	"	44	"	Sulphide of Iron.

$39 + 16 = 55.$

K S	"	55	"	Sulphide of Potassium.

In order to express two or more equivalents of a binary, we place a figure immediately before the symbol, like an algebraic coefficient. *A figure so placed always multiplies the whole binary.*

3 S O_3 stands for 3 equiv. or 120 parts of Sulphuric Acid.

5 Pb O   "   5   "   560   "   Oxide of Lead.

Ternary Compounds. — The symbol of a ternary compound is formed by writing together the symbols of the two binaries of which it consists, separated by a comma, taking care to place the most electro-positive binary, the base, first. If the salt is composed of more than one equivalent of either base or acid, then the number of equivalents must be indicated by

2 *

coefficients. The ternary symbol thus obtained always stands for one equivalent of the compound, and the weight of this equivalent is evidently the sum of the weights of the equiva‐ lents of the elements entering into it.

$1 + 8 + 16 + 24 = 49.$

$H O, S O_3$ stands for 1 equiv. or 49 parts of Sulphate of Water (common Sulphuric Acid),

$1 + 8 + 14 + 40 = 63.$

$H O, N O_5$ stands for 1 equiv. or 63 parts of Nitrate of Water (common Nitric Acid).

$39 + 8 + 6 + 16 = 69.$

$K O, C O_2$ stands for 1 equiv. or 69 parts of Carbonate of Potassa.

$23 + 8 + 16 + 24 = 71.$

$Na O, S O_3$ stands for 1 equiv. or 71 parts of Sulphate of Soda.

$39 + 8 + 27 + 24 = 98.$

$K O, Cr O_3$ stands for 1 equiv. or 98 parts of Neutral Chromate of Potassa.

$39 + 8 + 2 (27 + 24) = 149.$

$K O, 2 Cr O_3$ stands for 1 equiv. or 149 parts of Bichromate of Potassa.

$39 + 8 + 3 (27 + 24) = 200.$

$K O, 3 Cr O_3$ stands for 1 equiv. or 200 parts of Trichromate of Potassa.

$104 + 8 + 27 + 24 = 163.$

$Pb O, Cr O_3$ stands for 1 equiv. or 163 parts of Neutral Chro‐ mate of Lead. -

$2 (104 + 8) + 27 + 24 = 275.$

$2 Pb O, Cr O_3$ stands for 1 equiv. or 275 parts of Basic Chromate of Lead.

In order to express two or more equivalents of a ternary, we enclose the symbol in parentheses, and place before the whole the required figure. Thus,

3 (NaO, SO_3) stands for three equivalents of Sulphate of Soda.

5 ($2 PbO, CrO_3$) stands for five equivalents of Basic Chromate of Lead.

Neutral Salts. — The larger number of inorganic acids combine most readily with one equivalent of base, and the salts so formed will be called neutral salts. If the salts contain more equivalents of acid or base than one, they are called acid or basic salts respectively. There are, however, some acids which combine most readily with two or three equivalents of base, in the same way that the others combine with one. Such acids are called bibasic or tribasic acids, in order to distinguish them from the rest, which are frequently called monobasic. Of bibasic and tribasic acids, the most important in inorganic chemistry is Phosphoric Acid. This is known in three different conditions. In the first of these it is monobasic, in the second bibasic, and in the third tribasic, the last being the ordinary condition. The three conditions are designated by the symbol $_aPO_5$; $_bPO_5$; $_cPO_5$. The acid $_cPO_5$ forms neutral salts when combined with three equivalents of base, the acid $_bPO_5$ when combined with two, and the acid $_aPO_5$ when combined with only one. There are three compounds of the acid and water corresponding to the three conditions, which are represented in symbols by $HO, _aPO_5$; $2HO, _bPO_5$; $3HO, _cPO_5$. In these compounds we can substitute for the equivalents of water equivalents of other bases, either in whole or in part, forming such compounds as $NaO, _aPO_5$; $2NaO, _bPO_5$; $3NaO, _cPO_5$; $[HO, 2NaO]_cPO_5$; $[2HO, NaO]_cPO_5$. The equivalents of water may even be replaced by different bases, as in the compounds $[NaO, PbO]_bPO_5$; $[HO, NaO,$

K O], $_c$P O$_5$; [K O, 2 Mg O] $_c$P O$_5$. All the above are symbols of neutral salts.

As protoxide bases combine most readily with one equivalent
of acid, so sesquioxide bases combine most readily with three
equivalents. A neutral salt of a sesquioxide base is therefore
one which contains for every equivalent of base three equivalents of a monobasic acid, or one equivalent of a tribasic acid,
and for every two equivalents of base three equivalents of a
bibasic acid. Hence, $Fe_2 O_3 , 3 S O_3$; $2 Fe_2 O_3 , 3_b P O_5$;
$Fe_2 O_3 , {}_cP O_5$ are all symbols of neutral salts. On the other
hand, $Fe_2 O_3 , S O_3$; $4 Al_2 O_3 , 3_b P O_5$; $2 Al_2 O_3 , {}_cP O_5$ are symbols of basic salts. It will be noticed, on examining the above
symbols, that neutral salts of monobasic acids contain as many
equivalents of acid as there are equivalents of oxygen in the
base, and neutral salts of bibasic and tribasic acids one half
and one third as many, respectively. This rule must be kept
in mind when writing the symbols of salts.

Compound Radicals. — There is a large class of substances
which, although compound, nevertheless act in chemical changes
exactly as if they were simple, frequently replacing the elements themselves. Such substances are termed compound
radicals. Many of these radicals have, like the elements,
received arbitrary names, such as Cyanogen, Ammonium,
Ethyle, Acetyle, &c. ; and, moreover, the first letter or letters
of these names are frequently used as their symbols. It is
best, however, to write out the symbols of the elements forming these compounds, and enclose the whole in brackets; thus,
[N H$_4$] stands for Ammonium, [C$_2$ N] for Cyanogen, [C$_4$ H$_5$]
for Ethyle, [C$_4$ H$_3$] for Acetyle. The oxides of the compound

radicals, like those of the elements, may be divided into acids and bases, and their symbols are written exactly like those of other binaries. Thus,

$[C_4 H_3] O_3$ is the symbol of Acetic Acid.

$[N H_4] O$ "   "   Oxide of Ammonium.

$[C_4 H_5] O$ "   "   Oxide of Ethyle.

So, also, with the salts.

$[N H_4] O, [C_4 H_3] O_3$ is the symbol of Acetate of Ammonia.

$[C_4 H_5] O, [C_4 H_3] O_3$ "   "   Acetate of Oxide of Ethyle.

Water of Crystallization. — Besides the water of constitution, which frequently forms a part or the whole of the base of a salt, most salts combine with water as a whole. This water is held in combination by a comparatively feeble affinity, and may be generally driven off by exposing the salt to the temperature of 100° C., and sometimes escapes at the ordinary temperature of the air, the crystals of the salt in all cases falling into powder. Its presence is essential to the crystalline condition of many salts, and hence the name Water of Crystallization. The presence of water of crystallization in a salt is expressed in symbols, by writing after the symbol of the salt, and separated from it by a period, the number of equivalents of water. Thus,

$Fe O, S O_3 . 7 H O$ is the symbol of Crystallized Sulphate of the Oxide of Iron (Green Vitriol).

$H O, 2 Na O, {}_c P O_5 . 24 H O$ is the symbol of Crystallized Phosphate of Soda.

The same salt, when crystallized at different temperatures, not unfrequently combines with different amounts of water of crystallization, the less amounts corresponding to the higher tem-

peratures.   Thus, the Sulphate of Manganese may be crystallized with three different amounts of water of crystallization.

Mn O, S O$_3$. 7 H O when crystallized below 6° Centigrade.

Mn O, S O$_3$. 5 H O   "   "   between 7° and 20°.

Mn O, S O$_3$. 4 H O   "   "   between 20° and 30°.

The crystalline forms of these three compounds are entirely
different from each other, proving that the form depends, in
part at least, on the amount of water which the salt contains.

The symbols of other ternary compounds are written like
those of the oxygen salts, and therefore require no further
explanation.   Below are a few of these symbols, together with
those of the corresponding oxygen salts, which may serve as
examples.

K O, C O$_2$ = Carbonate of Oxide of Potassium.

K S, C S$_2$ = Sulphocarbonate of Sulphide of Potassium.

K O, As O$_3$ = Arsenite of Oxide of Potassium.

K S, As S$_3$ = Sulphoarsenite of Sulphide of Potassium.

3 Na Cl, Sb Cl$_3$ = Double Chloride of Antimony and Sodium.

[N H$_4$] Cl, Pt Cl$_2$ = Double Chloride of Platinum and Potassium.

K I, Pt I$_2$ = Double Iodide of Platinum and Potassium.

Quaternaries. — The symbols of the double salts are formed
by writing together the symbols of the two salts of which they
consist, separated by a period.   Thus,

K O, S O$_3$. Mg O, S O$_3$. 6 H O = Double Sulphate of Magnesia and Potassa.

K O, S O$_3$. Al$_2$ O$_3$, 3 S O$_3$. 24 H O = Double Sulphate of Alumina and Potassa (Alum).

When the salts contain water of crystallization, the amount of
this water expressed in equivalents is written after the symbol
of the salt, as already explained.

CHEMICAL REACTIONS.

THE various chemical changes to which all matter is more or less liable are termed, in the language of chemistry, *reactions*, and the agents which cause these changes, *reagents*. In every chemical reaction we must distinguish between the substances which are involved in the change and those which are produced by it. The first will be termed the *factors*, and the last the *products*, of the reaction. As matter is indestructible, it follows that *The sum of the weights of the products of any reaction must always be equal to the sum of the weights of the factors.* This statement seems at first sight to be contradicted by experience, since wood and many other combustible substances are apparently *consumed* by burning. In all such cases, however, the apparent annihilation of the substance arises from the fact that the products of the change are invisible gases; and when these are collected, their weight is found to be equal, not only to that of the substance, but also, in addition, to the weight of the oxygen from the air consumed in the process. As the products and factors of every chemical change must be equal, it follows that *A chemical reaction may always be repre-*

sented in an equation by writing the symbols of the factors in the first member, and those of the products in the second. The reaction of sulphuric acid on common salt may be represented by the following equation:

$$23 + 35 \quad + \quad 1 + 8 + 16 + 24 \ = \ 23 + 8 + 16 + 24 \quad + \quad 1 + 35 = 107.$$
$$Na\ Cl + H\ O,\ S\ O_3 = Na\ O,\ S\ O_3 + \mathbf{H\ Cl}.$$

The correctness of this may be proved by adding together the equivalents of both sides, when the sums will be found to be equal. In like manner, the reaction of a solution of common phosphate of soda on a solution of chloride of calcium may be represented by the equation

$$1 + 8 + 2\ (23 + 8) + 32 + 40 \ + \ 3\ (20 + 35) =$$
$$H\ O,\ 2\ Na\ O,\ _cP\ O_5 + 3\ Ca\ Cl + Aq^* =$$
$$3\ (20 + 8) + 32 + 40 \ + \ 2\ (23 + 35) \ + \ 1 + 35 = 308.$$
$$3\ Ca\ O,\ _cP\ O_5 + 2\ Na\ Cl + H\ Cl + Aq.$$

So, also, the reaction of hydrochloric acid on chalk, which may be proved like the other two:

$$Ca\ O,\ C\ O_2 + H\ Cl + Aq = Ca\ Cl + H\ O + Aq + \mathbf{C\ O_2}.$$

Although the equation is the most concise, and therefore in most cases the best form of representing chemical reactions, it is nevertheless frequently advantageous, in studying complicated changes, to adopt a more graphic method, by which the various steps of the process may be indicated. The reactions represented by the preceding equations may be written thus: —

---

* The symbol Aq, for Aqua, merely indicates the condition of solution, and is not to be regarded in adding up the equivalents in order to prove the equation.

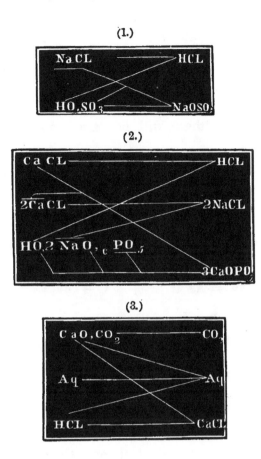

(1.)

NaCL   HCL

HO,SO$_3$ NaOSO$_3$

(2.)

CaCL   HCL

2CaCL   2NaCL

HO,2NaO,$_c$PO$_5$

3CaOPO$_4$

(3.)

CaO,CO$_2$ CO$_2$

Aq   Aq

HCL   CaCL

Chemical reactions may be classed under three divisions.

First, those reactions in which a compound is decomposed, and divides into simpler compounds or into elements. E. g. when oxide of mercury is heated, it is decomposed into oxygen gas and metallic mercury. Thus,

$$Hg\ O = Hg + \mathbf{O}.$$

Again, when Chlorate of Potassa is heated, it is resolved into oxygen gas and chloride of potassium. Thus,

$$K\ O,\ Cl\ O_5 = K\ Cl + 6\ \mathbf{O}.$$

3

So, also, when sulphate of lead is heated, it is resolved into anhydrous sulphuric acid and oxide of lead. Thus,

$$Pb\,O, S\,O_3 = Pb\,O + \mathbf{S\,O_3}.$$

Such reactions as these will be called *analytical*, and the process *analysis*.

Second, those reactions in which the elements are united to form compounds, or compounds of a lower order to form those of a higher. E. g. when hydrogen and carbon burn in the air, they combine with oxygen to form water or carbonic acid. Thus,

$$\mathbf{H + O = H\,O}; \qquad C + \mathbf{O_2} = \mathbf{C\,O_2}.$$

Again, when anhydrous sulphuric acid combines with lime to form sulphate of lime. Thus,

$$Ca\,O + S\,O_3 = Ca\,O, S\,O_3.$$

Reactions like these will be called *synthetical*, and the process *synthesis*.

Third, those reactions in which one element displaces another. E. G. when sodium takes the place of hydrogen in water, or zinc the place of hydrogen in dilute sulphuric acid. Thus,

$$H\,O + Na = Na\,O + \mathbf{H}.$$
$$Zn + H\,O, S\,O_3 + Aq = Zn\,O, S\,O_3 + Aq + \mathbf{H}.$$

This division includes also those reactions in which there is a mutual interchange of elements between two compounds. E. g. when a solution of chloride of barium is added to a solution of sulphate of soda, the sodium and barium change places, and we have formed an insoluble precipitate of sulphate of baryta and chloride of sodium (common salt), which remains in solution. Thus,

$$Ba\ Cl + Na\ O, S\ O_3 + Aq = Ba\ O, S\ O_3 + Na\ Cl + Aq.$$

Reactions like these will be called *metathetical*, and the process *metathesis*.*

Of the three classes of chemical reactions, the last is by far the most important; indeed, the larger number of reactions described in an elementary treatise on chemistry are examples of metathesis. All metathetical reactions can be illustrated very elegantly with the aid of mechanical diagrams, as follows : —

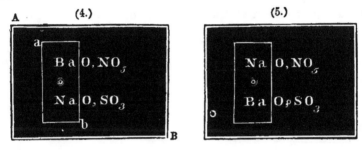

They are easily made by printing with stencils on the larger piece of pasteboard, A B, Fig. 4, the symbols of the elements not disturbed in the reaction, and on the smaller piece, a b, the symbols of the elements which exchange places. The smaller piece having been fastened to the larger by means of an eyelet at O, the reaction is represented by merely turning it half round. (See Fig. 5.) If the symbols of the interchanging elements are not symmetrical on all sides, it is of course necessary to make them reversible, by printing each on a separate small square of pasteboard, fastened by an eyelet to the

* From the Greek μετατίθημι, to displace or to transpose.

top or bottom of the revolving piece a b, since otherwise the letters would be inverted when the diagram is turned. This method of illustration may, with a little ingenuity, be extended to some of the most complicated cases of chemical change.

The most important condition of chemical action is, that the particles of the substances involved in the change should be indued with freedom of motion. This condition is generally fulfilled, both in nature and in our laboratories, by bringing the substances together in solution, either in water or in some other fluid. When substances are brought together in solution, there are two circumstances which, more than any others, determine the nature and extent of the resulting change.

First, *If by an interchange of analogous elements an insoluble compound may be formed, this compound always separates from the fluid as a precipitate.* As this circumstance is by far the most important of all in determining chemical reactions, it requires full illustration.

$$\left. \begin{array}{l} Ca\ O,\ N\ O_5 \\ Na\ O,\ C\ O_2 \end{array} \right\} + Aq = \left. \begin{array}{l} Na\ O,\ NO_5 \\ Ca\ O,\ C\ O_2 \end{array} \right\} + Aq.$$

$$\left. \begin{array}{l} Cu\ O,\ S\ O_3 \\ H\ S \end{array} \right\} + Aq = \left. \begin{array}{l} H\ O,\ S\ O_3 \\ Cu\ S \end{array} \right\} + Aq.$$

$$\left. \begin{array}{l} Ba\ O,\ N\ O_5 \\ H\ O,\ S\ O_3 \end{array} \right\} + Aq = \left. \begin{array}{l} H\ O,\ N\ O_5 \\ Ba\ O,\ S\ O_3 \end{array} \right\} Aq.$$

In these examples, and in general throughout the volume, the symbols of the substances, when in solution, are printed in italic letters, and the solid precipitate in Roman letters. The symbol Aq, as already stated, stands for an indefinite amount of water, in which the substances are supposed to be dissolved. It is obvious, from the above examples, that, in order to ascertain whether two salts will react on each other, when brought together in solution, so as to form a precipitate, it is only necessary to write the symbol of one under that of the other, and interchange the symbols of the metallic elements. If either compound whose symbols are thus formed is insoluble in the menstruum present, a reaction will take place, and the insoluble compound will be precipitated. At the end of the volume will be found a table, reprinted from the English edition of Fresenius's Qualitative Analysis, by means of which the student can easily ascertain from inspection the solubility of any of the more frequently occurring binary compounds or salts, and thus will be able to solve the following problems.

Problem 1. If chloride of barium and sulphate of soda are mixed together in solution, will there be a reaction; and if so, what will be formed?

Problem 2. If chloride of sodium and nitrate of silver are mixed together in solution, will there be a reaction, &c.?

Problem 3. If sulphide of hydrogen and nitrate of lead are mixed together in solution, will there be a reaction?

Problem 4. If sulphide of hydrogen and sulphate of zinc are mixed together in solution, will there be a reaction?

In solving the last problem, it must be noticed that the fluid which would result from a reaction would be a weak acid, in which many substances are soluble which would be insoluble in pure water, as may be seen from the table.

Problem 5. If chloride of sodium and sulphate of copper are mixed together in solution, will there be a reaction?

Problem 6. If sulphuric acid and borate of soda are mixed together in solution, will there be a reaction?

It will be found that, by an interchange of metallic elements in the last two examples, no insoluble compound will be formed, and hence the conclusion follows from our data, that there will be no precipitate. We must not, however, conclude from this that there will be no reaction, since, as can easily be seen, it does not necessarily follow, because the possible formation of an insoluble compound always determines a reaction, that the reverse is equally true, and that no reaction can take place unless an insoluble compound is formed. Indeed, in the last two examples, we are able, from incidental phenomena, to determine satisfactorily that a change does result; thus, in Problem 5, when the solutions are mixed, the blue color of sulphate of copper changes into the green color of chloride of copper; and in Problem 6, if sulphuric acid is not added in excess, the claret color to which blue litmus-paper turns in the mixed solution proves that it is boracic acid, and not sulphuric acid, which is in a free state; nevertheless, in most similar cases it is impossible to determine, unless an insoluble compound is formed, whether any reaction has taken place.

Second. The circumstance which, next to insolubility, is most important in determining metathetical reactions, is volatility, and it may be laid down as a general principle, that, *If by an interchange of analogous elements between two substances in solution, a substance can be formed, which is volatile at the temperature at which the experiment is conducted, such an interchange always takes place, and the volatile product is set free.* In order to illustrate this principle, a few examples may be adduced.

1. If diluted sulphuric acid is poured upon granulated zinc, a brisk evolution of hydrogen gas ensues, and sulphate of oxide of zinc is retained in solution. Thus,

$$\left.\begin{array}{l} \text{Zn} \\ H\,O, S\,O_3 + Aq \end{array}\right\} = \left\{\begin{array}{l} \mathbf{H} \\ Zn\,O, S\,O_3 + Aq. \end{array}\right.$$

In this example, and those that follow, the volatile or gaseous products are always printed with a full-face type.

2. If diluted sulphuric acid is poured upon protosulphide of iron, sulphide of hydrogen gas escapes, and sulphate of protoxide of iron remains in solution. Thus,

$$\left.\begin{array}{l} \text{Fe S} \\ H\,O, S\,O_3 + Aq \end{array}\right\} = \left\{\begin{array}{l} \mathbf{H\,S} \\ Fe\,O, S\,O_3 + Aq. \end{array}\right.$$

It will be noticed from the last two reactions, that it is not essential that more than one of the factors of the reaction should be fluid, or in solution.

3. If strong sulphuric acid is poured upon common salt, and the mixture slightly heated, chlorohydric acid gas is evolved, and bisulphate of soda remains dissolved in the excess of sulphuric acid. Thus,

$$\left.\begin{array}{l} Na\,Cl \\ H\,O, S\,O_3 \,.\, H\,O, S\,O_3 \end{array}\right\} = \left\{\begin{array}{l} \mathbf{H\,Cl} \\ Na\,O, S\,O_3 \,.\, H\,O, S\,O_3. \end{array}\right.$$

4. If strong sulphuric acid is poured upon nitre, and the temperature of the mixture slightly elevated, the vapor of nitric acid is given off, and bisulphate of potassa is formed. Thus,

$$\left.\begin{array}{l} K\,O, N\,O_5 \\ H\,O, S\,O_3 \,.\, H\,O, S\,O_3 \end{array}\right\} = \left\{\begin{array}{l} \mathbf{H\,O, N\,O_5} \\ K\,O, S\,O_3 \,.\, H\,O, S\,O_3. \end{array}\right.$$

5. If diluted nitric acid is poured upon chalk, or any other analogous carbonate, carbonic acid gas is set free, and a salt of nitric acid formed. Thus,

$$\left.\begin{array}{l} Ca\,O, C\,O_2 \\ H\,O, N\,O_5 + Aq \end{array}\right\} = \left\{\begin{array}{l} H\,O + \mathbf{C\,O_2} \\ Ca\,O, N\,O_5 + Aq. \end{array}\right.$$

It is very frequently the case that two, or even all of the three, classes of chemical reactions are combined, and going on simultaneously, in a single experiment. In the last example, for instance, the metathesis is succeeded by an analysis of one of the products, owing to the want of affinity between CO_2 and HO. Again, the reaction of nitric acid on copper is an example where all three varieties of reactions are combined. Three equivalents of copper react on four equivalents of nitric acid, and the reaction may be conveniently studied in two parts. In the first part, the copper is oxidized by one of the equivalents of the acid; here we have analysis accompanied by synthesis; and in the second part the copper changes place with the hydrogen of the acid, a case of metathesis.

1st. $3 Cu + HO, NO_5 = 3 (CuO, HO) + \mathbf{NO_2}$.

2d. $\left. \begin{matrix} HO, NO_5 \\ CuO, HO \end{matrix} \right\} + Aq = \left\{ \begin{matrix} CuO, NO_5 \\ HO + HO \end{matrix} \right\} + Aq.$

The whole reaction combined may be expressed thus:

$$3 Cu + 4 (HO, NO_5) + Aq = 3 (CuO, NO_5) + Aq + \mathbf{NO_2}.$$

Such mixed processes are very common in all complex cases of chemical change.

Stochiometrical Problems. — The chemical symbols enable us not only to represent chemical changes, but also to calculate exactly the amounts of the substances required in any given process, as well as the amounts of the products which it will yield. The method of making such calculation can best be illustrated by examples. In these examples the weights and measures will be given according to the French decimal system, which is now very generally used in chemical laboratories, and which, on account of the very simple relation between the units of measure and of weight, greatly facilitates stochiometrical calculations. The French measures and weights can, when required, be very easily reduced to the English standards, by means of a table at the end of the volume.

Problem 1. We have given 10 kilogrammes of common salt, and it is required to calculate how much chlorohydric acid gas can be obtained from it by treating with sulphuric acid. The reaction is

$$\overset{23+35.5\ =\ 58.5.}{Na\,Cl} + H\,O, S\,O_3 + Aq = Na\,O, S\,O_3 + Aq + \overset{1+35.5\ =\ 36.5.}{\mathbf{H}\,\mathbf{Cl}}.$$

Hence one equivalent, or 58.5 parts, of common salt, will yield one equivalent, or 36.5 parts, of chlorohydric acid gas. Therefore the amount which 10 kilogrammes will yield can be calculated from the proportion

$$\overset{53.5}{Na\,Cl} : \overset{36.5}{\mathbf{H}\,\mathbf{Cl}} = 10 : x = 6.239 \text{ kilogrammes, Ans.}$$

Problem 2. It is required to calculate how much chlorine gas can be obtained from chlorohydric acid with 6 grammes of hyperoxide of manganese. The equation representing the reaction is

$$\overset{28.6+16}{Mn\,O_2} + 2\,\overset{44.6.}{H\,Cl} + Aq = Mn\,Cl + 2\,H\,O + Aq + \overset{35.5.}{\mathbf{Cl}}.$$

Hence the amount of hyperoxide of manganese represented by $Mn\,O_2$, or 44.6 parts, yields an amount of chlorine gas represented by Cl, or 35.5 parts, and we have the proportion,

$$\overset{44.6}{Mn\,O_2} : \overset{35.5}{\mathbf{Cl}} = 6 : x = 4.775 \text{ grammes.}$$

Problem 3. It is required to calculate how much sulphuric

acid and nitre must be used to make 250 grammes of the strongest nitric acid.

$$\overset{101.2}{KO,NO_5} + \overset{98}{2(HO,SO_3)} = KO,SO_3 . HO,SO_3 + \overset{63}{\mathbf{HO,NO_5}}.$$

Henee,

$$\overset{63}{HO,NO_5} : \overset{98}{2(HO,SO_3)} = 250 : x.$$

$$\overset{63}{HO,NO_5} : \overset{101.2}{KO,NO_5} = 250 : x.$$

From the above examples we can deduce the following general rule for calculating from the amount of any given factor of a chemical reaction the amount of the products, or the reverse. *Express the reaction in an equation · make then the proportion, As the symbol of the substance given is to the symbol of the substance required, so is the amount of the substance given to x, the amount of the substance required; reduce the symbols to numbers, and calculate the value of x.*

On account of the very great lightness, the amount of a gas is very much more frequently estimated by measure than by weight. At the end of the volume a table will be found giving the weight in grammes of one thousand cubic centimetres, or one litre, of each of the most important gases. With the aid of this table such problems as the following may be solved.

Problem 4. How much chlorate of potassa must be used to obtain one litre of oxygen gas? One litre of oxygen gas weighs 1.43 grammes.

$$KO, ClO_5 = KCl + 6\,\mathbf{O}.$$

$$\overset{48}{6\,O} : \overset{122.7}{KO, ClO_5} = 1.43 : x.$$

Problem 5. How much zinc and sulphuric acid must be used to obtain 4 litres of hydrogen gas? Four litres of hydrogen weigh 0.357 grammes.

$$Zn + HO, SO_3 + Aq = ZnO, SO_3 + Aq + \mathbf{H}.$$

$$\overset{1}{\mathbf{H}} : \overset{32.6}{Zn} = 0.357 : x = \text{amount of zinc required.}$$

$$\overset{1}{\mathbf{H}} : \overset{49}{HO, SO_3} = 0.357 : x = \text{amount of sulphuric acid required.}$$

Problem 6. If ten grammes of water are decomposed by galvanism, how large a volume of mixed gases will they give?

$$HO = \mathbf{H} + \mathbf{O}.$$

$$\overset{9}{HO} : \overset{1}{\mathbf{H}} = 10 : x = \tfrac{10}{9} \text{ grammes of hydrogen.}$$

$$\overset{9}{HO} : \overset{8}{\mathbf{O}} = 10 : x = \tfrac{80}{9} \text{ grammes of oxygen.}$$

$\tfrac{10}{9}$ grammes of hydrogen occupy 12.429 cubic centimetres.

$\tfrac{80}{9}$ grammes of oxygen occupy   6.214   "   "

The mixed gases occupy   18.644   "   "

And hence * water, when decomposed into its elements, expands 1864 times.

The use of chemical symbols, both in expressing chemical reactions and in stochiometrical calculations, having been explained, they will be used in the following pages to illustrate the text of Stöckhardt's Elements of Chemistry. The reactions described in that work are represented in the form of equations, the first member containing always the factors, and

---

* It must be remembered that one cubic centimetre of water weighs one gramme.

the second, the products of the process. It remains for the student to work out the reaction, and represent it in the manner explained on page 25. The equivalents of water which are set free or formed during a reaction, are not generally indicated, but are merged in the general symbol Aq; and the student will frequently be obliged to supply these equivalents in order to work out the reaction. The symbols of solids are printed in Roman letters, those of fluids in italics, and those of gases in full-faced type. When, however, the solid or gas is dissolved in water, the symbol is printed in italics, followed by the general symbol Aq, or aq. Aq always stands for an indefinite and large amount of water; aq, for an indefinite but small amount of water. The color of a substance, especially of precipitates, is frequently printed above its symbol. The headings and figures, or letters at the side of the page, refer to the sections of the above-mentioned book.

In order still further to illustrate the subject, a large number of problems have been added to the reactions, which the student is expected to solve. The method of solving these problems can, in most cases, be deduced from the explanations already given, and from the sections of the text-book; in all other cases the method is explained under the problem. Throughout the following pages, all gases and vapors are supposed to be measured at the temperature of $0°$ and when the barometer stands at 76 centimetres, unless some other temperature or barometric pressure is expressly stated. All specific gravities are referred to water at its maximum density (at $4°$). The temperatures are all given on the Centigrade scale, and the weights and measures according to the French decimal system.

REACTIONS AND PROBLEMS.

4

WEIGHING AND MEASURING.

10. *Problems on French System of Weights and Measures.*

1. Reduce by means of the table at the end of the book, —
a. 30 inches to fractions of a metre.
b. 76 centimetres to English inches.
c. 36 feet to metres.
d. 10 metres to feet and inches.

2. Reduce by means of the table at the end of the book, —
a. 8 lbs. 6 oz. to grammes.
b. 7640 grammes to English apothecaries' weight.
c. 45 grains to grammes.

3. What is the diameter, and what is the circumference, of the globe in French measure?

4. What is the distance from Dunkirk in France to Barcelona in Spain? The latitude of Dunkirk = 51° 3', that of Barcelona = 41° 22', and the two places are on the same meridian.

5. Were our globe composed entirely of water at its greatest density, what would be its weight in kilogrammes?

6. What is the weight of one cubic decimetre of water?

7. Reduce by means of the table at the end of the book, —
a. 4 pints to litres and cubic centimetres.
b. 5 gallons to litres and cubic centimetres.

c. 5 litres to English measure.

d. 4 cubic centimetres to English measure.

11 – 17.   *Problems on Specific Gravity.*

1. The specific gravity of iron = 7.84. What is the weight of 1 cubic centimetre, 4 cubic centimetres, &c. of the metal in grammes?

2. The specific gravity of alcohol = 0.81. What is the weight of one litre in grammes? of 45 cubic centimetres, &c.?

3. The specific gravity of sulphuric acid = 1.85. If you wish to use in a chemical experiment 250 grammes, how much must you measure out?

4. Knowing the specific gravity of any given substance, how can you calculate the weight corresponding to any given measure, or the measure corresponding to any given weight? Give a general algebraic formula for the purpose, representing specific gravity, weight, and volume, by Sp. Gr., W., and V.

5. Determine the Sp. Gr. of absolute alcohol from the following data : —

	Grammes.
Weight of bottle empty,	4.326
" " filled with water at 4°,	19.654
" " " alcohol at 0°,	16.741

6. Determine the Sp. Gr. of lead shot from the following data : —

	Grammes.
Weight of bottle empty,	4.326
" " " filled with water at 4°,	19.654
" " shot,	15.456
" " bottle, shot, and water,	33.766

7. Determine the Sp. Gr. of iron from the following data : —

	Grammes.
Weight of iron in air,	3.92
" " " water,	3.42

8. Determine the Sp. Gr. of copper from

	Grammes.
Weight of copper in air,	10.000
" " " water,	8.864

9. Determine the Sp. Gr. of ash wood from

	Grammes.
Weight of wood in air,	25.350
" a copper sinker,	11.000
" wood and sinker under water,	5.100

10. How much bulk must a hollow vessel of copper fill, weighing one kilogramme, which will just float in water?

11. How much bulk must a hollow vessel of iron occupy, weighing ten kilogrammes, which sinks one half in water?

12. An alloy of gold and silver weighs ten kilogrammes in the air, and 9.735 kilogrammes in water. What are the proportions of gold and silver? Sp. Gr. of gold $= 19.2$, of silver $= 10.5$.

Solution. — In the French system the volume of a solid in cubic centimetres equals its weight in grammes divided by its Sp. Gr., or $V = \frac{w}{Sp. Gr.}$. Since one cubic centimetre of water weighs one gramme, the volume of a solid in cubic centimetres is equal to the weight of water it displaces in grammes. Hence the weight of water displaced $= \frac{w}{Sp. Gr.}$. Put, then, $x =$ weight of gold in alloy, $10 - x$ will equal weight of silver. $\frac{x}{19.2} =$ weight of water displaced by gold, and $\frac{10-x}{10.5} =$ weight of water displaced by silver. Hence $\frac{x}{19.2} + \frac{10-x}{10.5} = 0.265$.

13. An alloy of copper and silver weighs 37 kilogrammes in the air, and loses 3.666 kilogrammes when weighed in water. What are the proportions of silver and copper?

22.   *Problems on Expansion of Liquids and Gases.*

1. If 34.562 cubic centimetres of mercury at 0° are heated to 100°, what increase of volume do they undergo, and what is the increased volume?

Solution. — The small fraction of its volume by which one c. c. of a liquid

4 *

or gas increases when heated from 0° to 1°, is called the *coefficient of expansion* of that liquid or gas. The coefficient of expansion of mercury, for example, = 0.00018, that is, one c. c. of mercury at 0° becomes 1.00018 c. c. at 1°. If we assume that the expansion is proportional to the temperature, then one c. c. at 0° becomes 1.00036 at 2°, 1.0009 at 5°, and 1.018 at 100°. Hence, 34.562 c. c. of mercury would become, at 100°, (34.562 × 1.018) c. c. To make this solution general, let k = coefficient of expansion; then $(1 + k)$ = increased volume of one c. c. when heated from 0° to 1°, and $(1 + t\,k)$ = increased volume of one c. c. when heated from 0° to $t°$, and $V (1 + t\,k)$ = increased volume of V c. c. when heated from 0° to $t°$; representing by V' the increased volume, we have

$$V' = V (1 + t° k),$$

from which the increased volume of any liquid or gas may be calculated when the volume at 0°, the coefficient of expansion, and the temperature are known.

It is not true, as we have assumed, that liquids expand twice as much for two degrees, three times as much for three degrees, &c., as they do for one; but, on the contrary, the rate of expansion slowly increases with the temperature. For example, one c. c. of mercury at 0° becomes 1.000179 at 1°, but one c. c. at 300° becomes 1.000194 at 331°. This difference of rate, however, is so small, that we can neglect it, except in the most refined experiments, more especially if we use not the coefficient observed at any particular temperature, but a mean coefficient obtained by observing the total amount of expansion between 0° and 100°, and then dividing the result by 100, by which we averrge the error. Again, experiments on the expansion of fluids are commonly conducted in glass vessels, which expand themselves by heat, and therefore cause the expansion to appear less than it is. If they expanded as much as the fluid, it is evident that the fluid would not appear to expand at all. They in fact expand much less than the fluids, but, nevertheless, sufficiently to make a material difference between the *absolute expansion* of a fluid, and its *apparent expansion* in glass vessels. The mean absolute coefficient of mercury between 0° and 100° is 1.000181. The apparent expansion in glass between 0° and 100° is 1.000156. In Table IV., at the end of the volume, will be found the mean coefficients of expansion in glass of some of the more important fluids. The rate of expansion of water varies so rapidly and so anomalously, that no use should be made of its coefficient except in experiments extending over 100°. When it is required to determine the amount of expansion between narrower limits, use should be made of Table V., which gives the volume to which one cubic centimetre of water at 0° or at 4° increases, when heated to the temperatures at the side of the table. It also gives the Sp. Gr. of water at different temperatures, when either water at 0° or at 4° is taken as the unit.

2. What will be the volume of 5.346 c. c. of water at 0°, when heated to 100°?

3. What will be the volume of 250 c. c. of oil of turpentine, at 0°, when heated to 50°?

4. What will be the volume of 35 c. c. of water at 0°, when heated to 4°, to 25°, to 40°, and to 84° ?

17. *Problems on reducing Specific Gravities to the Standard Temperature.*

The specific gravities of both liquids and solids are supposed to be referred to water at 4°, its greatest density; but in practice we always use water at a much higher temperature, and it becomes therefore necessary to reduce the results to 4°, or in other words, to calculate what would have been the result had the temperature been 4° during the experiment. Hence an important class of problems like the following : —

1. The Sp. Gr. of zinc was found to be 7.1582 when the temperature of the water was 15°. What would have been the Sp. Gr. at 4° ?

Solution. — Sp. Gr. of water at 4° : Sp. Gr. at 15° = Sp. Gr. of zinc at 15° : Sp. Gr. at 4°; or 1 : 0.9992647 = 7.1582 : x.

2. The Sp. Gr. of antimony was found to be 6.681 when the temperature of the water was 15°. What would have been the Sp. Gr. at 4° ? Ans. 6.677.

3. The Sp. Gr. of an alloy of zinc and antimony was found from the following data : —:

	Grammes.
Weight of the alloy,	4.4106
" Sp. Gr. bottle,	9.0560
" " " full of water at 4°,	19.0910
" bottle, alloy, and water at 14°.6,	22.8035
	Ans. 6.375.

4. Find the Sp. Gr. of metallic zinc from the following data : —

	Grammes.
Weight of the zinc,	12.4145
" bottle,	9.0560
" " full of water at 18°,	19.0790
" zinc and water at 12°.4,	29.7663
	Ans. 7.153.

24. *Problems on reducing Centigrade Degrees to Fahrenheit, and the reverse.*

1. What do —40°, —20°, —15°, —9°, 0°, 4°, 10°, 13°, 15°, 80°, 100°, 150° Cent. correspond to on the Fahrenheit scale?

Rule. — Double the number of degrees, subtract one tenth of the whole, and add 32 if the degrees are above 0°, or subtract 32 if they are below.

2. What do —40°, —32°, —18°, —7°, 0°, 10°, 32°, 42°, 50°, 70°, 90°, 100°, 150°, 212°, 300°, 450° Fahr. correspond to on the Centigrade scale?

Rule. — Subtract 32 if above 0°, or add 32 if below, add one ninth to the result, and divide by two.

27.   *Problems on Expansion of Solids.*

1. What will be the length of a rod of iron 424.56 metres long at 0°, when heated to 20°?

Solution. — The small fraction of its length by which a rod of iron, or of any other solid, one metre long, expands when heated from 0° to 1°, is called the coefficient of linear expansion of the solid. A bar of iron one metre long at 0° becomes 1.0000122 at 1°, and the small fraction 0.0000122 is the coefficient of linear expansion of iron. Representing this coefficient by k, we have for the new length of the rod, at 1°, $l' = l (1 + k)$, and at $t°$, $l' = l(1 + t k)$. We may assume, in the case of solids, that the expansion is proportional to the temperature, especially if we deduce our coefficient from experiments made between 0° and 100°, as described under examples on the expansion of fluids. Substituting for l, k, and t the values given in the problem, we have $l' = 424.56 (1 + 20 \times 0.0000122)$. The coefficients of linear expansion for a number of solids are given in Table IV. at the end of the volume.

2. What will be the length of a rod of copper 2.365 metres long, at 0°, when heated to 100°?

3. What will be the length of a rod of silver 0.760 metres long, at 12°, when heated to 20°?

Solution. — Denoting by l the unknown length of the rod at 0°, by l' the known length at $t°$, and l'' the required length at $t'°$, we have as above,

$$l' = l (1 + t k), \quad \text{and} \quad l'' = l (1 + t' k).$$

By combining these, we óbtain

$$l'' = l\left(\frac{1+t'k}{1+t\,k}\right) = l\,[1 + k\,(t' - t) + \&c.].$$

All the terms of the quotient may be neglected after the first, because they contain powers of the already very small fraction k. Substituting the values in the above equation, we get

$$l'' = 0.760\,[1 + 0.000019\,(20 - 12)].$$

4. One of the large iron tubes of the Britannia Bridge over the Menai Straits is 143.253 metres long. What increase of length does it undergo between 8° and 20° Centigrade?

5. What is the increased capacity of a globe of glass which holds exactly one litre, at 0°, when heated to 250°?

Solution. — If we consider for a moment the glass globe as forming the outside shell of a solid globe of glass, it is evident that the increased capacity of the globe will be equal to the increased volume of this solid globe, which we have supposed for a moment to fill the interior. The problem, therefore, resolves itself into calculating the amount of cubic expansion of a solid glass globe having a volume equal to one litre, or 1000 c. c. The coefficient of linear expansion of glass is given in the table as 0.00001. The coefficient of cubic expansion is always three times * as great as the linear expansion ; in the case of glass, therefore, it is 0.00003. By this is meant that, if a rod of glass one metre long at 0° becomes 1.00001 long at 1°, then a cube of glass of one cubic centimetre at 0° becomes 1.00003 c. c. at 1°. Using, then, the equation

$$V' = V\,(1 + t\,k),$$

we obtain by substitution

$$V' = 1000\,(1 + 250 \times 0.00003) = 1007.5 \text{ c. c.}$$

6. What is the increased capacity of a globe of glass which holds exactly 495 c. c. at 15°, when heated to 300°?

---

* Each edge of a cube of glass one centimetre long at 0°, would become $(1 + k)$ c. m. long at 1°. The increased volume of the cube would be equal to $(1 + k)^3 = 1 + 3k + 3k^2 + k^3$; but as k is an exceedingly small fraction, k^2 and k^3 may be neglected in comparison, so that a cube of glass of one c. c. at 0° becomes $(1 + 3k)$ at 1°, which proves that the amount of cubic expansion is three times as great as the linear.

NON-METALLIC ELEMENTS, OR METALLOIDS.

Oxygen (O).

56. $\overset{\text{Red.}}{\text{Hg}} O = Hg + O.$

1. How much oxygen can be obtained by heating 108 grammes, 250 grammes, 25 grammes, or 5 grammes, of red oxide of mercury?

2. How many cubic centimetres of oxygen can be obtained from the amounts of oxide of mercury given in the last example?

3. How much oxide of mercury would be required to yield one litre of oxygen by the process of 1?

4. How much mercury would remain after the experiment of last example?

59. $K O, Cl O_5 = K Cl + 6 O.$

1. How much oxygen can be obtained from 1 kilogramme, 50 grammes, or 5 grammes, of chlorate of potassa?

2. How much chloride of potassium would remain after the oxygen in the last examples had been driven off?

3. How much chlorate of potassa would be required to make one litre of oxygen?

63. $C + 2O = CO_2$.

1. How much carbonic acid gas will be formed by burning 5 grammes of carbon?

2. How many cubic centimetres of carbonic acid will be formed by burning 5 grammes of carbon?

3. How much oxygen will be consumed in the last two examples?

4. How many cubic centimetres of carbonic acid will be formed from one litre of oxygen, by burning in it carbon?

64. $S + 2O = SO_2$.

1. How much sulphurous acid gas will be formed by burning 10 grammes of sulphur?

2. How much oxygen will be consumed in the last example?

3. How many cubic centimetres of sulphurous acid will be formed by burning sulphur in one litre of oxygen?

4. Assuming that one litre of oxygen yields exactly one litre of sulphurous acid, what is the Sp. Gr. of SO_2 gas?

65. $P + 5O = PO_5$.

1. How much phosphoric acid can be formed from 48 grammes of phosphorus?

2. How much phosphorus will exactly consume one litre of oxygen?

3. How much phosphorus is required in order to make 250 grammes of phosphoric acid?

67. $Na + \bullet = Na\,O.$

    1. How much oxide of sodium can be made from 28.75 grammes of sodium?

    2. How much oxide of sodium can be made with one litre of oxygen? and how much sodium will be consumed in the experiment?

68. $3\,Fe + 4\,\bullet = Fe\,O,\,Fe_2\,O_3.$

71. $2\,Na\,O + P\,O_5 + Aq = H\,O,\,2\,Na\,O,\,P\,O_5 + Aq.$

    The crystallized salt $= [H\,O,\,2\,Na\,O]\,P\,O_5\,.\,24\,H\,O.$

72. $Ca\,O + C\,O_2 + Aq = Ca\,O,\,C\,O_2 + Aq.$

73. $Ca\,O + S\,O_2 + Aq = Ca\,O,\,S\,O_2 + Aq$

    1. How much lime would be required to neutralize 20 grammes of sulphurous acid?

    2. How much lime would be required to neutralize the sulphurous acid obtained by burning 5 grammes of sulphur?

    3. How much lime would be required to neutralize one litre of sulphurous acid gas?

79. $3\,Mn\,O_2 = Mn\,O,\,Mn_2\,O_3 + 2\,\bullet.$

    1. How much oxygen can be obtained from one kilogramme of hyperoxide of manganese by heating?

    2. How much of the oxide must be used in order to obtain 30 grammes of oxygen?

$$Mn\,O_2 + H\,O,\,S\,O_3 = Mn\,O,\,S\,O_3 + H\,O + \bullet.$$

    1. How much oxygen can be obtained from one kilogr. of hyperoxide of manganese by the last process? and how much sulphuric acid will be required to decompose it?

    2. What per cent of the whole amount of oxygen in the mineral is obtained by the two processes? and by how much does the second exceed the first?

Hydrogen (H).

81. $Na + Aq = Na\,O, H\,O + Aq + \mathbf{H}.$

1. How much hydrogen will be set free by 5.25 grammes of sodium? How many cubic centimetres will be set free?

82. $3\,Fe + 4\,\mathbf{H}\,\mathbf{O} = Fe\,O, Fe_2\,O_3 + 4\,\mathbf{H}.$

1. How much hydrogen would be obtained in the last experiment by the decomposition of 39 grammes of water?

2. By how much would the weight of the tube increase in last example?

83. $Zn + H\,O, S\,O_3 + Aq = Zn\,O, S\,O_3 + Aq + \mathbf{H}.$

1. How much sulphuric acid and how much zinc must be used in order to make 2 grammes of hydrogen? How much in order to make one litre?

2. How much hydrogen can be obtained with one kilogramme of zinc?

3. How much with one kilogramme of sulphuric acid?

Decomposed by Galvanism.

55. $H\,O = \mathbf{H} + \mathbf{O}.$

1. How many cubic centimetres of hydrogen would be obtained by the decomposition of one cubic centimetre (one gramme) of water? How many cubic centimetres of oxygen? How many of mixed gases?

85.   A solid immersed in a liquid or a gas is buoyed up by a force equal to the weight of the liquid or gas which it displaces. The excess of the buoyancy over its own weight is called its ascensional force.

1. What would be the ascensional force of a small balloon filled with one litre of hydrogen gas, when the balloon itself weighs five centigrammes?

5

2. What would be the ascensional force of a spherical balloon seven metres in diameter, two thirds filled with hydrogen, when the balloon and attachments weigh twenty kilogrammes?

87. $\mathbf{H} + \mathbf{O} = HO.$

1. How much water would be formed by burning one thousand litres of hydrogen? and how much oxygen would be consumed in the process?

2. How much vapor of water would be formed in Example 1?

93.   *Problems on the Barometer.*

1. When the surface of the column of mercury in a barometer stands at 76 centimetres above the mercury in the basin, with what weight is the atmosphere pressing on every square centimetre of surface? Sp. Gr. of mercury $= 13.596$.

2. To what difference of pressure does a difference of one centimetre in the barometric column correspond?

3. When the water barometer stands at ten metres, what is the pressure of the air if the temperature is 4°?

4. How high would an alcohol barometer, and how high a sulphuric-acid barometer, stand under the same circumstances, disregarding in each case the tension of the vapor? Sp. Gr. of alcohol $= 0.8095$; Sp. Gr. of sulphuric acid $= 1.85$.

94. *Problems on the Compression and Expansion of Gases.*

Mariotte's Law. — It is an established principle of science that *The volume of a given weight of gas is inversely as the pressure to which it is exposed ;* that is, the greater the pressure, the smaller is the volume; and the less the pressure, the larger is the volume. This may be illustrated by an India-rubber bag holding one litre of air, or of any other gas. This is exposed to a pressure, under the ordinary conditions of the atmosphere, of a little over one kilogramme on every square centimetre of surface. If this pressure is doubled, the volume of the bag will be reduced to one half; if trebled, to one third, &c. On the other hand, if the pressure is reduced to one half, the volume will double; if to one third, the volume will treble, &c.* The principle is expressed in mathematical language by the proportion

$$H' : H = V : V'; \qquad (1.)$$

where H and H' are the heights of the barometer which measure the pressure to which the gas is exposed under the two conditions of volume V and V'.

Since the density of a given weight of gas is inversely as the volume, or $D' : D = V : V'$, it follows that

$$H' : H = D' : D, \qquad (2.)$$

or the density of a gas is proportional to the pressure to which it is exposed. Moreover, since the weight of a given volume of gas is proportional to its density, and its density, as just proved, proportional to the pressure, it follows that *The weight of a given volume of gas is directly as the pressure to which it is exposed,* or

$$H' : H = W' : W. \qquad (3.)$$

These three proportions are very important, and will be constantly referred to in the following pages. The student must be careful to notice that in (1) the weight of gas is supposed to be constant and the volume to vary, and in (2) the volume is supposed to be constant and the weight to vary.

---

* We suppose the bag to have no elasticity.

The variations in the pressure of the atmosphere, amounting at times to one tenth of the whole, necessarily cause equally great changes in the volume of gases which are the objects of chemical experiment. Hence, in order to compare together different volumes of gas, it is essential that they should have been measured when exposed to the same pressure. A standard pressure has therefore been agreed upon, that measured by 76 centimetres of mercury, to which the volume of gases measured under any other pressure should be reduced. Hence a number of problems like the following : —

1. A volume of hydrogen gas was measured and found to be equal to 250 c. c. The height of the barometer, observed at the same time, was 74.2 centim. What would have been the volume if observed when the barometer stood at 76 centim. ?

Solution. — Proportion (1) gives, by substituting the data of the problem, 74.2 : 76 = 250 : V′ = Ans.

2. A volume of nitrogen gas measured 756 c. c. when the barometer stood at 77.4 centim. What would it have measured if the barometer had stood at 76 centim. ?

3. A volume of air standing in a bell-glass over a mercury pneumatic trough measured 568 c. c. The barometer at the time stood at 75.4 centim., and the surface of the mercury in the bell was found, by measurement, to be 6.5 centim. above the surface of the mercury in the trough. What would have been the volume had the air been exposed to the pressure of 76 centim. ?

Solution. — It can easily be seen, that the pressure of the air on the surface of the mercury in the pneumatic trough, measured by the height of the barometer at the time (75.4 centim.), was balanced first by the column of mercury in the bell, and secondly by the tension * of the confined air. Hence, the pressure * to which the air was exposed was

---

* The tension of a gas is the force with which it tends to expand, and, when the gas is at rest, must evidently be exactly equal to the pressure to which it is exposed.

equal to the height of the barometer less the height of the mercury in the bell, or 75.4 — 6.5 = 68.9 centim. We have then the proportion 68.9 : 76 = 568 : V′ = Ans.

4. A volume of air standing in a tall bell-glass over a mercury pneumatic trough measured 78 c. c. The barometer at the time stood at 74.6 centim., and the mercury in the bell at 57.4 centim. above the mercury in the trough. What would have been the volume had the pressure been 76 centim.?

5. What would be the answers to the last two problems, had the pneumatic trough been filled with water instead of mercury?

97.   *Problems on Expansion of Gases by Heat.*

1. What will be the volume of 250 cubic centimetres of air at 0° when heated to 300°?

Solution. — It has been found by Regnault and others, that the permanent gases expand so nearly equally for the same increase of temperature, that the differences may be entirely disregarded except in the most refined investigations; and it has also been found, that their rate of expansion does not materially vary from the lowest to the highest temperatures at which experiments have been made. The coefficient of expansion for air, as determined by Regnault, is equal to 0.00366, and we can therefore calculate the volume, V′, of a gas at any temperature, from its volume, V, at 0°, by means of the equation, already explained,

$$V′ = V (1 + t \times 0.00366); \quad\quad (1.)$$

or if we know the volume V′ for a temperature t, we can calculate the Volume V″ for another temperature $t′$, by means of the equation

$$V″ = V′ (1 + (t′ — t) 0.00366). \quad\quad (2.)$$

By transposing, we can obtain from equation (1)

$$V = \frac{V′}{1 + t \times 0.00366}. \quad\quad (3.)$$

As the volume of a gas varies very considerably with the temperature, it is important, in comparing together different measurements, that we should adopt a standard temperature, as we have adopted a standard pressure. The temperature which has been agreed upon is 0°; but as it would be inconvenient, and often impossible, to make our measurements at this temperature, it becomes necessary to calculate, by means of equa-

5 *

tion (3), from a volume V′ measured at $t°$, what would be the volume at 0°. This is called technically reducing the volume to 0°. There can be obtained also from equations (1) and (2) the equations

$$t = \frac{V' - V}{V \times 0.00366}, \quad (4.) \qquad \text{and} \quad t' = t + \frac{V'' - V'}{V' \times 0.00366}, \quad (5.)$$

by means of which we can calculate the change of temperature when we know the change of volume. Representing the coefficient of expansion in the above formulæ by k, we can obtain, by transposing and reducing the equation,

$$k = \frac{V' - V'}{t \, V}, \quad (6.) \qquad \text{and} \quad k = \frac{V'' - V'}{V' \, (t' - t)}. \quad (7.)$$

From these we can calculate the coefficient of expansion when we know the volume of a gas at two different temperatures.

2. A volume of gas measured 560 c. c. at 15°. What would it measure at 95° ?

3. A glass globe holding 450 c. c. of air at 0° was heated to 300°. At this temperature the neck was hermetically sealed, and the globe cooled again to 0°. The neck was then opened under mercury, and the air remaining in the globe passed up into a graduated jar, and measured. How much was it found to measure ?

Solution. — By substituting the values for V and t in the equation V′ = V $(1 + t \times 0.00003)$, we obtain the increased capacity of the globe, and of course the number of cubic centimetres of expanded air which it contains at 300°. It is then only necessary to substitute this value for V′ and 300 for t, in equation (3), in order to find what will be the volume of this expanded air when cooled again to 0°.

4. What is the weight of air contained in an open glass globe of 250 c. c. capacity, at the temperature of 20°, and when the barometer stands at 74 centimetres ?

Solution. — In order to make the solution general, we will represent the capacity of the globe, the temperature, and the height of the barometer, by V, t, and H, respectively. One cubic centimetre of air at 0°, and when the barometer stands at 76 centimetres, weighs 0.00129 grammes. To find what one cubic centimetre would weigh when the barometer stands at H centimetres, we make use of proportion (3), page 51:

H : H′ = W : W′; or 76 : H = 0.00129 : W′;

whence W = 0.00129 . $\frac{H}{76}$, the weight of one cubic centimetre at 0°, and under a pressure of H centimetres. To find what one cubic centimetre would weigh at t^o, it must be remembered that one cubic centimetre at 0° becomes $(1 + t\,0.00366)$ c. c. at t^o; therefore, at t^o and at H centimetres of the barometer, $(1 + t\,0.00366)$ c. c. weigh $0.00129 \cdot \frac{H}{76}$ grammes. By equating these two terms we obtain $(1 + t\,0.00366) = 0.00129 \cdot \frac{H}{76}$, whence $1 = 0.00129 \cdot \frac{1}{1 + t\,0.00366} \cdot \frac{H}{76}$, the weight of one cubic centimetre at t^o and under a pressure of H centimetres. The weight of V cubic centimetres, w, is evidently

$$w = 0.00129 \; V \cdot \frac{1}{1 + t\,0.00366} \cdot \frac{H}{76}. \qquad (8.)$$

5. What is the weight of air contained in an open glass globe of 560 cubic centimetres' capacity at $0°$, at the temperature of $300°$, and under a pressure of 77 centimetres?

Solution. — In the solution of the last example we neglected the change of capacity of the glass globe due to the change of temperature. This causes no sensible error when the change of temperature is small, but when, as in the present problem, the change of temperature is quite large, the change of capacity of the globe must be considered. If the capacity is V c. c. at $0°$, it becomes at t^o V $(1 + t\,0.00003)$. Introducing this value for V into the equations of the last section, we obtain

$$w = 0.00129 \; V \; (1 + t\,0.00003) \cdot \frac{1}{1 + t\,0.00366} \cdot \frac{H}{76}. \qquad (9.)$$

99. *Problems on Specific Gravity of Vapors.*

General Solution. — The specific gravity of a vapor is its weight compared with the weight of the same volume of air under the same conditions of temperature and pressure. To find, then, the specific gravity of a vapor, we must ascertain the weight of a known volume, V, at a known temperature, t, and under a known pressure, H, and divide this by the weight of the same volume of air at the same temperature, and under the same pressure. The method may best be explained by an example. Suppose, then, that we wish to ascertain the specific gravity of alcohol vapor. We take a light glass globe having a capacity of from 400 to 500 c. c., and

Fig. 6.

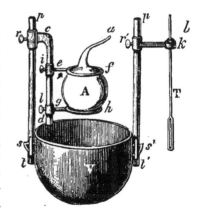

draw the neck out in the flame of a blast lamp, so as to leave only a fine opening, as shown in the figure at a. The first step is now to ascertain the weight of the glass globe when completely exhausted of air. As this cannot readily be done directly, we weigh the globe full of air, and then subtract the weight of the air, ascertained by calculation from the capacity of the globe, and from the temperature and pressure of the air, by means of equation (8). Call the weight of the globe and air W, and the weight of the air w, then $W - w$ is the weight of the globe exhausted of air. The second step is to ascertain the weight of the globe filled with alcohol vapor at a known temperature, and under a known pressure. For this purpose we introduce into the globe a few grammes of pure alcohol, and mount it on the support represented in Fig. 6. By loosening the screw, r, we next sink the balloon beneath the oil contained in the iron vessel, V, and secure it in this position. We now slowly raise the temperature of the oil to between 300° and 400°, which we observe by means of the thermometer, B. The alcohol changes to vapor and drives out the air, which, with the excess of vapor, escapes at a. When the bath has attained the requisite temperature, we close the opening a, by suddenly melting the end of the tube at a with a mouth blow-pipe, and as nearly as possible at the same moment observe the temperature of the bath and the height of the barometer. We have now the globe filled with alcohol vapor at a known temperature, and under a known pressure. Since it is hermetically sealed, its weight cannot change, and we can therefore allow it to cool, clean it, and weigh it at our leisure. This will give us the weight of the globe filled with alcohol vapor at a known temperature, t', and under a known pressure, H'. Call this weight W'. The weight of the vapor is $W' - W + w$. The third step is to ascertain the weight of the same volume of air at the same temperature and under the same pressure. This can easily be found by calculation from equation (9). The last step is to find the capacity of the globe, which, although we have supposed it known, is not actually ascertained experimentally until the end of the process. For this purpose we break off the tip of the tube (a), under mercury, which, if the experiment has been carefully conducted, rushes in and fills the globe completely. We then empty this mercury into a carefully graduated glass cylinder, and read off the volume. We find then the specific gravity by dividing the weight of the vapor by the weight of the air. The formulæ for the calculation are then

Weight of the globe and air,   W.

"   "   air, $w = 0.00129 \text{ V} \cdot \dfrac{1}{1 + t\, 0.00366} \cdot \dfrac{H}{76}.$

"   "   globe exhausted of air, $W - w$.

"   "   " filled with vapor at a temperature t' and under a pressure H',   W'.

Weight of the vapor, $\qquad$ $W' - W + w.$

" " air at t' and under a pressure H', $\bigg\} = 0.00129\, V\,(1 + t'\,0.00003) \cdot \dfrac{1}{1 + t'\,0.00366} \cdot \dfrac{H}{76}.$

$$ \text{Sp. Gr.} = \frac{W' - W + w}{0.00129\, V\,(1 + t'\,0.00003) \cdot \dfrac{1}{1 + t\,0.00366} \cdot \dfrac{H}{76}}. $$

1. Ascertain the Sp. Gr. of alcohol vapor from the following data: —

Weight of glass globe,	W	50.8039 grammes.
Height of barometer,	H	74.754 centim.
Temperature,	t	18°
Weight of globe and vapor,	W'	50.8245 grammes.
Height of barometer,	H'	74.764 centim.
Temperature,	t'	167°
Volume,	V	351.5 cubic centim.

Ans. 1.5795.

2. Ascertain the Sp. Gr. of camphor vapor from the following data: —

Weight of glass globe,	W	50.1342 grammes.
Height of barometer,	H	74.2 centim.
Temperature,	t	13°.5
Weight of globe and vapor,	W'	50.8422 grammes.
Height of barometer,	H'	74.2 centim.
Temperature,	t	244°
Volume,	V	295 cubic centim.

Ans. 5.298.

Carbon (C).

109. $C + 2\,O = C\,O_2.$

$2\,Hg\,O + C = 2\,Hg + C\,O_2.$

1. How many grammes and how many cubic centimetres of carbonic acid gas are formed by burning 10 grammes of charcoal?

2. How many grammes and how many cubic centimetres of oxygen are consumed in the process?

3. Assuming that the volume of carbonic acid gas generated during combustion is exactly equal to the volume of oxygen gas consumed, what is the Sp. Gr. of carbonic acid gas?

4. How much oxide of mercury is required to burn up 5.672 grammes of charcoal?

110. $C + O = CO.$ $C + CO_2 = 2CO.$

1. How many grammes and how many cubic centimetres of oxide of carbon gas are formed by burning 10 grammes of charcoal?

2. How many grammes and how many cubic centimetres of oxygen are consumed in the process?

3. Ten cubic centimetres of oxygen yield how many cubic centimetres of carbonic acid gas, and how many of oxide of carbon gas? What expansion ·does oxygen undergo in combining with carbon to form oxide of carbon?

Spermaceti.
115. $C_{64} H_{64} O_4 + 188 O = 64 CO_2 + 64 HO.$

1. How many grammes of carbonic acid, and how many grammes of water, are formed by burning 10 grammes of spermaceti?

2. The carbonic acid and water given off by a burning spermaceti candle were carefully collected and weighed. The water weighed 0.564 grammes, the carbonic acid weighed 1.3786. How much of the candle was burned?

SECOND GROUP OF METALLOIDS.

Sulphur (S).

132. $Fe\,S + HO,\,SO_3 + Aq = Fe\,O,\,SO_3 + Aq + \mathbf{H\,S}.$

1. How much sulphide of hydrogen can be made from 15 grammes of sulphide of iron? How many cubic centimetres?

2. How much sulphide of iron, and how much sulphuric acid, is required to generate sufficient gas to saturate one litre of water?

$$HS + Aq + \mathbf{O} = S + Aq.$$

$$\mathbf{H\,S} + 3\,\mathbf{O} = \mathbf{H\,O} + \mathbf{S\,O_2}.$$

133. *a.* Pb $+ HS + Aq =$ $\overset{\text{Black.}}{\text{Pb S}} + Aq + \mathbf{H}.$

b. $\overset{\text{Yellow.}}{\text{Pb O}} + HS + Aq = \overset{\text{Black.}}{\text{Pb S}} + Aq.$

c. $Pb\,O,\,[C_4\,H_3]\,O_3 + HS + Aq = \overset{\text{Black.}}{\text{Pb S}} +$ $HO,\,[C_4\,H_3]\,O_3 + Aq.$

d. $Fe\,O,\,SO_3 + Ca\,O,\,HO + HS + Aq = \overset{\text{Black.}}{\text{Fe S}} +$ $Ca\,O,\,SO_3 + Aq.$

$Cu\,O,\,[C_4\,H_3]\,O_3 + HS + Aq = \overset{\text{Black.}}{\text{Cu S}} +$ $HO,\,[C_4\,H_3]\,O_3 + Aq.$

$Sb\,Cl_3 + 3\,HS + Aq = \overset{\text{Orange.}}{\text{Sb S}_3} + 3\,H\,Cl + Aq.$

$As\,Cl_3 + 3\,HS + Aq = \overset{\text{Yellow.}}{\text{As S}_3} + 3\,H\,Cl + Aq.$

$Zn\,Cl + Ca\,S + Aq = \overset{\text{White.}}{\text{Zn S}} + Ca\,Cl + Aq.$

Phosphorus (P).

140. $P + 3\,O = P\,O_3$ (by slow combustion).

$P + 5\,O = P\,O_5$ (by rapid combustion).

144.   Burnt bones consist chiefly of $3\,Ca\,O,\ _cP\,O_5$.

$3\,Ca\,O,\ _cP\,O_5 + 2\,(H\,O,\,S'O_3) + aq = 2\,(Ca\,O,\,S\,O_3)$
$+ [2\,H\,O,\ Ca\,O]\ _cP\,O_5 + aq.$

Heated to a red heat.

$[2\,H\,O,\ Ca\,O],\ _cP\,O_5 + Aq + x\,C = Ca\,O,\ P\,O_5$
$+ x\,C + \mathbf{Aq} + 2\,\mathbf{C\,O} + 2\,\mathbf{H}.$

Heated intensely.

$3\,(Ca\,O,\ _aP\,O_5) + x\,C = 3\,Ca\,O,\ _aP\,O_5 + x\,C + 2\,\mathbf{P}$
$+ 10\,\mathbf{C\,O}.$

1. How much phosphorus can be manufactured from 20 kilogrammes of burnt bones, of which four fifths are phosphate of lime?

145. $4\,P + 3\,(Ca\,O,\,H\,O) = 3\,(Ca\,O,\,P\,O) + \mathbf{P\,H_3}.$

Besides the above reaction, there take place simultaneously the two following reactions, in the experiment described in the text-book.

$3\,P + 2\,(Ca\,O,\,H\,O) = 2\,(Ca\,O,\,P\,O) + \mathbf{P\,H_2}.$

$P + Ca\,O,\,H\,O = Ca\,O,\,P\,O + \mathbf{H}.$

THIRD GROUP OF METALLOIDS.

Chlorine (**Cl**).

150. $Mn\,O_2 + 2\,H\,Cl + aq = Mn\,Cl + aq + \mathbf{Cl}.$

1. How much chlorine gas can be obtained from 2.467

grammes of chlorohydric acid gas? How many cubic centimetres?

2. How much chlorine can be obtained from an undetermined amount of muriatic acid by means of 4.567 grammes of hyperoxide of manganese? How many cubic centimetres?

3. The hyperoxide of manganese of commerce is more or less adulterated. What per cent of $Mn\,O_2$ does an article contain, of which 10. grammes, when heated with strong muriatic acid, evolve 4.0135 grammes of chlorine?

4. How much chlorine can be obtained from 25 cubic centimetres of muriatic acid* of Sp. Gr. $= 1.16$? How many cubic centimetres?

5. In order to prepare one litre of chlorine gas how much hyperoxide of manganese, and how much muriatic acid, must be used? Calculate the amounts for pure $Mn\,O_2$ and $H\,Cl$ gas, and also when the oxide used contains only 70 per cent of pure $Mn\,O_2$, and when the liquid acid used has a Sp. Gr. $= 1.15$.

151. $Mn\,O_2 + 2\,Na\,Cl + 2\,(H\,O,\,S\,O_3) + aq = Mn\,Cl + 2\,(Na\,O,\,S\,O_3) + aq + \mathbf{Cl}.$

We might use one half as much common salt, but then we should find sulphate of manganese instead of chloride of manganese in solution. Thus,

$Mn\,O_2 + Na\,Cl + 2\,(H\,O,\,S\,O_3) + aq = Mn\,O,\,S\,O_3 + Na\,O,\,S\,O_3 + aq + \mathbf{Cl}.$

1. How much chlorine gas can be obtained by the last process from 34 kilogrammes of salt?

---

* See Table VI., which gives the per cent of $H\,Cl$ in the fluid acid of different specific gravities.

2. How many cubic centimetres of chlorine can be obtained from one cubic centimetre of rock salt? Sp. Gr. of salt = 2.15.

152. *f.* $2 (Fe\ O, S\ O_3) + H\ O, S\ O_3 + Cl + Aq = Fe_2\ O_3,$
$3\ S\ O_3 + H\ Cl + Aq.$

$6 (Fe\ O, S\ O_3) + 3\ Cl + Aq = 2 (Fe_2\ O_3, 3\ S\ O_3)$
$+ Fe_2\ Cl_3 + Aq.$

g. $Au + 3\ Cl + Aq = Au\ Cl_3 + Aq.$

ACIDS.

___

Nitrogen and Oxygen.

Nitric Acid (HO, NO_5).

159. $KO, NO_5 + 2(HO, SO_3) = KO, SO_3 . HO, SO_3$
$+ \mathbf{HO, NO_5}.$

$NaO, NO_5 + 2(HO, SO_3) = NaO, SO_3 . HO, SO_3$
$+ \mathbf{HO, NO_5}.$

1. How much nitric acid can be made from 250 kilogrammes of potash nitre, and how much sulphuric acid must be used in the process?

2. How much more nitric acid will the same weight of soda nitre yield?

3. How much nitric acid, containing 40 per cent. of NO_5, can be made from 1700 kilogrammes of potash nitre?

4. How much soda nitre, and how much sulphuric acid, and how much water, must be used to make 450 kilogrammes of nitric acid, which shall contain 60 per cent. of pure acid?

Ammonia.

160. *c.* $[N H_4] O, H O + H O, N O_5 + Aq = [N H_4] O, N O_5 + Aq.$

d. $Pb O + H O, N O_5 + Aq = Pb O, N O_5 + Aq.$

1. How much nitric acid, of Sp. Gr. 1.14, is required to dissolve 20 kilogrammes of oxide of lead?

2. How much nitric acid, of Sp. Gr. 1.14, and how much oxide of lead, must be used to make 10 kilogrammes of nitrate of lead?

e. $3 Pb + 4 (H O, N O_5) + aq = 3 (Pb O, N O_5) + aq + \mathbf{N O_2}.$

$3 Cu + 4 (H O, N O_5) + aq = 3 (Cu O, N O_5) + aq + \mathbf{N O_2}.$

1. How much nitric acid, of Sp. Gr. 1.22, is required to dissolve 450 grammes of lead? How much nitrate of lead would be formed?

2. How much nitric acid, of Sp. Gr. 1.362, is required to dissolve 450 grammes of copper?

f. $3 P + 5 (H O, N O_5) + Aq = 3 (3 H O, _c P O_5) + Aq + 5 \mathbf{N O_2}.$

$S + H O, N O_5 + Aq = H O, S O_3 + Aq + \mathbf{N O_2}.$

Nitric Oxide ($\mathbf{N O}$).

$6 Fe\ Cl + K O, N O_5 + 4 H Cl + Aq = 3 Fe_2\ Cl_3 + K Cl + aq + \mathbf{N O_2}.$

Colorless.   Red.

$\mathbf{N O_2} + \mathbf{O_2} = \mathbf{N O_4}.$

1. How much $\mathbf{N O_2}$ can be obtained by dissolving 10 grammes of copper in nitric acid? How many cubic centimetres?

2. How much $N O_2$ can be obtained from 10 grammes of iron by the last reaction but one?

3. What volume of oxygen must be mixed with one litre of $N O_2$ in order to change it into $N O_4$?

Nitrous Oxide ($N O$).

When heated.

163. $[N H_4] O, N O_5 = 4 H O + 2 N O.$

Sp. Gr. of 1.14.

$4 Pb + 5 (H O, N O_5) + Aq = 4 (Pb O, N O_5) + Aq + N O.$

$Na O, S O_2 + Aq + N O_2 = Na O, S O_3 + Aq + N O.$

1. Ten grammes of nitrate of ammonia yield how many grammes and how many cubic centimetres of protoxide of nitrogen?

2. How much nitrate of ammonia must be used in order to make one litre of the gas?

3. One litre of $N O_2$ yields how many cubic centimetres of $N O$ by the third reaction of this section?

4. One litre of $N O$ gives, when decomposed, what volume of nitrogen?

Carbon and Oxygen.

Carbonic Acid ($C O_2$).

164. $Ca O, C O_2 + H O, N O_5 + Aq = Ca O, N O_5 + Aq + C O_2.$

$Ca O, C O_2 + H O, S O_3 + Aq = Ca O, S O_3 + Aq + C O_2.$

6 *

$$Ca\ O,\ N\ O_5\ +\ H\ O,\ S\ O_3\ +\ aq\ =\ Ca\ O,\ S\ O_3\ +$$
$$H\ O,\ N\ O_5\ +\ aq.$$

1. How much sulphuric acid and how much nitric acid* must be used to drive out all the carbonic acid from 25.462 grammes of chalk? How many grammes and how many cubic centimetres of gas would be obtained?

2. The specific gravity of Carrara marble is 2.716. How many cubic centimetres of carbonic acid gas does one cubic centimetre of the marble contain in a condensed state?

167. 1. Animals remove oxygen from the air, and return the whole as carbonic acid. Plants remove carbonic acid, and, having decomposed it, return the oxygen it contained. How does the volume of the oxygen in either case compare with that of the carbonic acid?

Sulphur and Oxygen.

Sulphuric Acid ($S\ O_3$).

169. $\mathbf{S\ O_2 + O}† = S\ O_3$.

1. How much anhydrous sulphuric acid is formed by the oxidation of 10 grammes of sulphurous acid? and how much oxygen is required in the process?

2. How much anhydrous sulphuric acid is formed by the oxidation of one litre of sulphurous acid gas? and how many cubic centimetres of oxygen must be mixed with it in the experiment?

$a.\ 2\ S\ O_3 + \mathbf{H\ O} = H\ O,\ 2\ S\ O_3$ (the Nordhausen Acid).

---

* When the strength of the nitric acid is not stated, monohydrated acid ($H\ O,\ N\ O_5$) is always intended.

† The two gases are mixed together and led over heated platinum sponge in a glass tube.

$S O_3 + H O = H O, S O_3$ (the common Acid).

170. Fe O, $S O_3$. 6 H O is the symbol of crystallized green vitriol.

When heated.

$$2 (Fe O, S O_3 . 6 H O) = Fe_2 O_3, S O_3 + 12 \mathbf{H O} + \mathbf{S O_2}.$$

By further heating.

$$Fe_2 O_3, S O_3 = Fe_2 O_3 + \mathbf{S O_3}.$$

By conducting the anhydrous acid fumes into $H O, S O_3$ we get $H O, 2 S O_3$.

1. How much anhydrous acid, and how much Nord-hausen, can be made by the above process from 20 kilo-grammes of green vitriol? If the Nordhausen acid has the specific gravity of 1.9, how many litres can be obtained from 20 kilogrammes of green vitriol?

171. $\mathbf{S O_2 + H O, N O_5 + Aq} = H O, S O_3 + Aq + \mathbf{N O_4}.$

1. How much $H O, S O_3$ will be formed from one gramme of $S O_2$? How much from one litre?

White.

$$Ba\,Cl + H O, S O_3 + Aq = Ba\,O, S O_3 + H\,Cl + Aq.$$

$$Ba\,O, N O_5 + H O, S O_3 + Aq = Ba\,O, S O_3 + H\,Cl + Aq.$$

1. Why must sulphuric acid or a soluble sulphate pro-duce a precipitate when added, in solution, to the solution of any salt of baryta?

2. The precipitate produced by adding an excess of Ba Cl to a solution of $H O, S O_3$ was collected, and weighed 4.567 grammes. How much sulphuric acid* was present in solution?

* When the name sulphuric acid is used, $H O, S O_3$ is always meant, unless otherwise specified.

3. The precipitate produced by adding an excess of HO, SO_3 to a solution of BaO, NO_5 weighed 5.942 grammes. How much BaO, NO_5 did the solution contain?

172. 1st stage. $S + O_2 = SO_2,$ and $KO, NO_5 + 2(HO, SO_3)$
$$= KO, SO_3 . HO, SO_3 + HO, NO_5.$$

2d stage. $SO_2 + HO, NO_5 = HO, SO_3 + NO_4.$

3d stage. $3 NO_4 + x HO = 2(HO, NO_5) + NO_2.$

4th stage. $\begin{cases} NO_2 + O_2 = NO_4. \\ 2 SO_2 + 2(HO, NO_5) = 2(HO, SO_3) \\ \quad + 2 NO_4. \end{cases}$

The last two stages are now repeated indefinitely, so long as there is a supply of sulphurous acid, oxygen, and steam, with the same amount of NO_4.

1. How much sulphuric acid may be made by the above process from 100 kilogrammes of sulphur? How many litres of acid having the Sp. Gr. 1.842? How many of acid of Sp. Gr. 1.734? How many cubic metres of oxygen must be used in the process? How many cubic metres of air must pass through the lead chamber, supposing all its oxygen to be removed?

173. $a.$ $HO, SO_3 + Aq = HO, SO + Aq.$

1. How much water must one kilogramme of the monohydrated acid withdraw from the air in order to reduce its Sp. Gr. to 1.398?

$f.$ $NaO, CO_2 + HO, SO_3 + Aq = NaO, SO_3 + Aq + CO_2.$

1. How much HO, SO_3 is required to exactly neutralize 5.645 grammes of anhydrous carbonate of soda? How much acid of the Sp. Gr. 1.306?

2. How much $Na\,O,\,C\,O_2$ must be dissolved in one litre of water so as to make a solution such that one cubic centimetre will exactly neutralize 0.01 of a gramme of $H\,O,\,S\,O_3$?

Yellow.   White.

$g.\ \mathrm{Pb}\,O + H\,O,\,S\,O_3 + Aq = \mathrm{Pb}\,O,\,S\,O_3 + Aq.$

Black.   Blue Solution.

$h.\ \mathrm{Cu}\,O + H\,O,\,S\,O_3 + Aq = Cu\,O,\,S\,O_3 + Aq,$

which, when evaporated, gives crystals of the composition

Blue Vitriol.

$\mathrm{Cu}\,O,\,S\,O_3\,.\,5\,H\,O.$

1. How much sulphate of lead can be made from twenty grammes of litharge? How much sulphuric acid must be used in the process?

2. How much crystallized Blue Vitriol can be made from one kilogramme of oxide of copper? How much sulphuric acid of Sp. Gr. 1.615 must be measured out for the process?

174. $\mathrm{Cu} + 2\,(H\,O,\,S\,O_3) = \mathrm{Cu}\,O,\,S\,O_3 + 2\,\mathbf{H\,O} + \mathbf{S\,O_2}.$

175. $\mathrm{C} + 2\,(H\,O,\,S\,O_3) = 2\,\mathbf{H\,O} + 2\,\mathbf{S\,O_2} + \mathbf{C\,O_2}.$

$Na\,O,\,C\,O_2 + S\,O_2 + Aq = Na\,O,\,S\,O_2 + Aq + \mathbf{C\,O_2}.$

1. How much sulphurous acid can be made from 4.562 grammes of sulphuric acid by means of copper? How much by means of charcoal? How much anhydrous sulphate of copper would be formed in the first case? How much carbonic acid in the second? How much carbonate of soda will the sulphurous acid in the two examples neutralize? What is the volume of $S\,O_2$, and what the volume of $C\,O_2$ evolved in the second case?

2. How much copper and how much sulphuric acid must be used to make one litre of sulphurous acid gas?

How much to make 500 grammes of anhydrous sulphite of soda?

3. By burning sulphur in one litre of oxygen, how much $S O_2$ gas is obtained? What is the Sp. Gr. of $S O_2$?

Phosphorus and Oxygen.

Phosphoric Acid ($P O_5$).

176. $P + 5\,\mathbf{O} = P O_5$ (white powder).

Colorless Fluid.

$3\,P + 5\,(H O,\ N O_5) + Aq = 3\,H O,\ _cP O_5 + Aq + 5\,\mathbf{N\,O_2}.$

For preparation of phosphoric acid from bones, see § 144.

At the ordinary temperature.

$3\,Ca\,O,\ _cP O_5 + 2\,(H O,\ S O_3) + aq = 2\,(Ca\,O,\ S O_3) + [2\,H O,\ Ca\,O]\,_cP O_5 + aq.$

Intensely heated.

$2\,(Ca\,O,\ S O_3) + [2\,H O,\ Ca O]\,_cP O_5 = 3\,Ca\,O\,_cP O_5 + 2\,(\mathbf{H\,O,\ S\,O_3}).$

Before ignition.

$3\,H O, P O_5 + 3\,([N H_4]\,O, H O) + 3\,(Ag\,O, N O_5) + Aq$

Yellow.

$= 3\,Ag\,O,\ _cP O_5 + 3\,([N H_4]\,O,\ N O_5) + Aq.$

After ignition.

$H O,\ P O_5 + [N H_4]\,O,\ H O + Ag\,O,\ N O_5 + Aq =$

White.

$Ag\,O,\ _aP O_5 + [N H_4]\,O,\ N O_5 + Aq.$

$3\,H O,\ _cP O_5 + 2\,(Mg\,O,\ S O_3) + 3\,[N H_4]\,O,\ H O + Aq = [N\,H_4]O,\ 2\,Mg\,O,\ _cP O_5 . 12\,H O + 2\,([N H_4]\,O,\ S O_3) + Aq.$

[N H_4] O, 2 Mg O, $_c$P O_5 . 12 H O when ignited resolves into 2 Mg O, $_b$P O_5 + **N H_3** + 13 **H O.**

1. How much P O_5 and how much 3 H O, $_c$P O_5 can be obtained from 16 grammes of phosphorus ?

2. By boiling one gramme of phosphorus in nitric acid until it dissolves, diluting, neutralizing with aqua ammonia, and precipitating with a solution of sulphate of magnesia, collecting and igniting the precipitate, how much will it be found to weigh ?

3. How much nitrate of silver is required to precipitate the phosphoric acid made from one gramme of phosphorus before ignition ?   How much after ignition ?

Cyanogen and Oxygen.

179. Cy O = Cyanic Acid, which is *monobasic.*
$Cy_2 O_2$ = Fulminic Acid, which is *bibasic.*
$Cy_3 O_3$ = Cyanuric Acid, which is *tribasic.*

Boron and Oxygen.

Boracic Acid (B O_3).

180. $Na\ O,\ 2\ B\ O_3\ +\ H\ Cl\ +\ Aq\ =\ 2\ (H\ O,\ B\ O_3)$
$+\ Na\ Cl + Aq.$

The crystallized boracic acid is H O, B O_3 . 2 H O.

Dried at 100° it becomes H O, 2 B O_3 . 2 H O.

At a red heat it loses its water and melts, and on cooling it hardens to a vitreous mass.

Silicon and Oxygen.

Silicic Acid ($Si\ O_3$).

$$Na\ O,\ Si\ O_3 + H\ Cl + Aq = H\ O,\ Si\ O_3 + Na\ Cl + Aq.$$

If the quantity of water is large, the hydrated silicic acid remains in solution. If the amount of water is small, it separates as gelatinous precipitate.

SECOND GROUP: HYDROGEN ACIDS, OR COMPOUNDS OF THE HALOGENS WITH HYDROGEN.

Chlorine and Hydrogen.

Chlorohydric Acid (H Cl).

185. $Na\ Cl + H\ O,\ S\ O_3 = Na\ O,\ S\ O_3 + \mathbf{H\ Cl}$; or

$$Na\ Cl + 2\,(H\ O,\ S\ O_3) = Na\ O,\ S\ O_3\ .\ H\ O,\ S\ O_3 + \mathbf{H\ Cl}.$$

Only one equivalent of $H\ O,\ S\ O_3$ is necessary to decompose one equivalent of salt; but then the last half of the **H Cl** can be driven off only at a temperature sufficiently high to melt glass, so that with these proportions the process cannot be conducted in glass vessels. If two equivalents of $H\ O,\ S\ O_3$ are used, the whole of the **H Cl** is expelled at a moderate temperature, and the process can then be conducted to its end in a glass flask or retort. Hence, in the manufactories, where the acid is generally generated in iron retorts, only one equivalent of $H\ O,\ S\ O_3$ is used, while in the laboratory, where glass vessels are employed in the process, two equivalents are taken to each equivalent of salt. The last we will assume to be the case in the following problems.

1. How much chlorohydric acid gas can be made from 4.562 grammes; from 25 kilogrammes; from 34.567 grammes of common salt?

2. How much sulphuric acid is required to decompose the above amounts of common salt? and how much bisulphate of soda is in each case formed?

3. How many cubic centimetres of $\mathbf{H\,Cl}$ can be made from 1 gramme; from 5.643 grammes of Na Cl?

4. How much salt and how much sulphuric acid are required in order to make one kilogramme, to make 5.463 grammes, and to make one litre, of $\mathbf{H\,Cl}$?

5. How much $\mathbf{H\,Cl}$ is contained in one litre of the liquid acid of Sp. Gr. 1.16, of Sp. Gr. 1.17, of Sp. Gr. 1.14?

6. How many cubic centimetres of gas are dissolved in one litre of the liquid acid of the above strengths?

7. How much Na Cl, and how much HO, SO_3, and how much water in the receiver, are required to make, —

a. 20 kilogrammes of liquid acid of Sp. Gr. 1.13?

b. 560.4 grammes of liquid acid of Sp. Gr. 1.18?

c. 4 litres of liquid acid of Sp. Gr. 1.16?

$\mathbf{H} + \mathbf{Cl} = \mathbf{H\,Cl}$.

1. One litre of hydrogen gas combines with what volume of chlorine gas? and what is the volume of hydrochloric acid gas formed?

186. a. $\mathrm{Fe} + H\,Cl + Aq = Fe\,Cl + Aq + \mathbf{H}$.

1. How much liquid acid, by weight and by measure, of Sp. Gr. 1.16 is required to dissolve 250 grammes of

7

iron? How much chloride of iron would be obtained, and how much hydrogen gas, by measure, evolved?

$$\text{Sn} + H\,Cl + Aq = Sn\ Cl + Aq + \mathbf{H}.$$

1. Solve the last problem, substituting tin for iron.

b. $Fe_2\,O_3,\, 3\,H\,O + 3\,H\,Cl + Aq = Fe_2\ Cl_3 + Aq.$

c. $2\,Fe\ Cl + Cl + Aq = Fe_2\ Cl_3 + Aq.$

d. $Na\ O,\ C\,O_2 + H\,Cl + Aq = Na\ Cl + Aq + \mathbf{C\,O_2}.$

e. $Ag\,O,\,N\,O_5 + H\,Cl + Aq = \text{Ag}\,Cl + H\,O,\,N\,O_5 + Aq.$

1. How much liquid acid of Sp. Gr. 1.16 is required to dissolve 4 grammes of iron-rust?

2. How many cubic centimetres of chlorine are required to convert one gramme of protochloride of iron into sesquichloride?

3. How much liquid acid of Sp. Gr. 1.13 is required to neutralize one gramme of carbonate of soda?

4. How much **H Cl** do 50 c. c. of a liquid contain which is exactly neutralized by one gramme of anhydrous carbonate of soda?

5. How much **H Cl** do 50 c. c. of a liquid contain which gives, with an excess of nitrate of silver, a precipitate weighing 5.643 grammes?

Aqua Regia.

$$H\,O,\,N\,O_5 + 3\,H\,Cl + aq = N\,O_2\,Cl_2 + Cl + aq.$$
$$\text{Au} + (N\,O_2\,Cl_2 + Cl + aq) = Au\ Cl_3 + aq + N\,O_2.$$

Bromohydric Acid (H Br). Iodohydric Acid (H I).

$$P Br_3 + 3 HO = PO_3 + 3 \textbf{H Br}.$$

$$P I_3 + 3 HO = PO_3 + 3 \textbf{H I}.$$

Hydrofluoric Acid (H Fl).

$$Ca Fl + HO, SO + aq = CaO, SO_3 + HFl + aq.$$

$$SiO_3 + 3 \textbf{H Fl} = 3 HO + \textbf{Si Fl}_3.$$

Tartaric Acid ($2 HO, C_8 H_4 O_{10}$).

194. $2 ([N H_4] O, HO) + 2 HO, C_8 H_4 O_{10} + Aq =$
$2 [NH_4] O, C_8 H_4 O_{10} + Aq.$

$2 (K O, C O_2) + 2 HO, C_8 H_4 O_{10} + Aq =$
$2 KO, C_8 H_4 O_{10} + Aq + 2 \textbf{C O}_2.$

$2 KO, C_8 H_4 O_{10} + HCl + Aq = HO, KO, C_8 H_4 O_{10}{}^*$
$+ KCl + Aq.$

$2 (CaO, CO_2) + 2 (HO, KO, C_8 H_4 O_{10}) + Aq =$
$2 CaO, C_4 H_8 O_{10} + 2 KO, C_8 H_4 O_{10} + Aq + 2 \textbf{CO}_2.$

$2 KO, C_8 H_4 O_{10} + 2 Ca Cl + Aq = 2 Ca O, C_8 H_4 O_{10}$
$+ 2 K Cl + Aq.$

$2 CaO, C_4 H_8 O_{10} + 2 (HO, SO_3) + Aq = 2 (CaO, SO_3)$
$+ 2 HO, C_4 H_8 O_{10} + Aq.$

---

* This salt is not absolutely insoluble, but only difficultly soluble in water, and hence is not completely deposited in this reaction.

1. How much KO, CO_2 is required to exactly neutralize 5.462 grammes of tartaric acid?

2. How much HCl is required to convert 4.678 grammes of $2KO, C_8H_4O_{10}$ into $HO, KO, C_8H_4O_{10}$?

3. From ten kilogrammes of cream of tartar how much tartaric acid can be made?

Oxalic Acid (HO, C_2O_3).

The above is the symbol of the acid dried at 100°. When crystallized, it contains two more equivalents of water, and corresponds to the formula $HO, C_2O_3 . 2HO$.

196. $HO, C_2O_3 . 2HO + x(HO, SO_3) = x(HO, SO_3)$
$+ 3HO + \mathbf{CO_2} + \mathbf{CO}.$

197. $b.$ $KO, CO_2 + HO, C_2O_3 + Aq = KO, C_2O_3 + Aq$
$+ \mathbf{CO_2}.$

$KO, C_2O_3 + HO, C_2O_3 + Aq = KO, 2C_2O_3 + Aq.$

$d.$ $CaO, SO_3 + HO, C_2O_3 + Aq = CaO, C_2O_3$
$+ HO, SO_3 + Aq.$

$CaO, SO_3 + [NH_4]O, C_2O_3 + Aq = CaO, C_2O_3$
$+ (NH_4)O, SO_3 + Aq.$

1. How much $\mathbf{CO_2}$ and how much $\mathbf{CO}$ will be obtained by decomposing five grammes of crystallized oxalic acid by sulphuric acid? How many cubic centimetres of each gas?

2. How much crystallized oxalic acid must be used to yield one litre of $\mathbf{CO_2}$ and one litre of $\mathbf{CO}$?

3. How much crystallized oxalic acid will exactly neutralize 1.456 grammes of carbonate of potassa?

Acetic Acid ($HO, [C_4 H_3] O_3$).

198. $PbO + HO, [C_4 H_3] O_3 + Aq = Pb O, [C_4 H_3] O_3 + Aq.$

Crystallized acetate of lead $= PbO, [C_4 H_3] O_3 . 3\ HO.$

$Pb O, (C_4 H_3) O_3 + HO, SO_3 + Aq = PbO, SO_3 + HO, [C_4 H_3] O_3 + Aq.$

7 *

LIGHT METALS.

FIRST GROUP: ALKALI METALS.

Potassium (K).

202. $KO, CO_2 + HO, [C_4H_3] O_3 + Aq = KO, [C_4H_3] O_3 + Aq + \mathbf{C O_2}.$

$KO, CO_2 + HO, SO_3 + Aq = KO, SO_3 + Aq + \mathbf{CO_2}.$

1. How much HO, SO_3 must be dissolved in water in order to make a litre of test acid such that one cubic centimetre will exactly neutralize one decigramme of KO, CO_2?

2. How much crystallized oxalic acid must be dissolved in water in order to make a litre of test acid such that one hundred cubic centimetres will exactly neutralize 6.92 grammes of KO, CO_2?

203. $KO, CO_2 + CaO, HO + Aq = CaO, CO_2 + KO, HO + Aq.$

Melted together.
204. d. $SiO_3 + KO, HO = KO, SiO_3 + \mathbf{H O}.$

Light Blue.
e. $CuO, SO_3 + KO, HO + Aq = CuO, HO + KO, SO_3 + Aq.$

205. $\overset{\text{Intensely heated.}}{KO, CO_2} + 2C = \mathbf{K} + 3\mathbf{CO}.$

206. $KO, CO_2 + 2(HO, SO_3) + aq = KO, SO_3 . HO, SO_3$
$+ aq + \mathbf{CO_2}.$

207. $KO, CO_2 + HO, NO_5 + aq = KO, NO_5 + aq$
$+ \mathbf{CO_2}.$

$\overset{\text{When heated.}}{a. \ KO, NO_5} = KO, NO_3 + \mathbf{O_2}.$

$b. \ KO, NO_5 + 3C + S = KS + 3\mathbf{CO_2} + \mathbf{N}.$

1. How many cubic centimetres of mixed gases are formed by the burning of one kilogramme of gunpowder, when measured at the standard temperature and pressure? Assuming that the temperature at the time is 1000°, what would be the volume of the gases the moment after the explosion.

2. Assuming that gunpowder occupies the same bulk as an equal weight of water, into how many times its own volume does it expand on burning? Calculate both for 0° and for 1000°.

3. Assuming that the temperature, the moment after explosion, is 1000°, what would be the pressure on the interior surface of a bomb of 20 centimetres internal diameter when exploded filled with gunpowder?

208. $a. \ KO, ClO_5 = KCl + 6\mathbf{O}.$

1. How much does the oxygen contained in chlorate of potassa expand when the salt is decomposed? Sp. Gr. of chlorate of potassa is 2 nearly.

2. How much mechanical force would be required to reduce oxygen gas to the same degree of condensation in which it exists in the salt?

$c. \ 3(KO, ClO_5) + 4(HO, SO_3) = 2(KO, SO_3 . HO, SO_3)$
$+ KO, ClO_7 + 2\mathbf{ClO_4}.$

$$yf \ KO, Cl\, O_5 + 6\, H\, Cl + Aq = K\, Cl + Aq + 2\, \mathbf{Cl}.$$

$$6\, (K\, O, H\, O) + 6\, Cl + Aq = KO, Cl\, O_5 + 5\, K\, Cl + Aq.$$

210. $KI + Mn\, O_2 + 2\, (H\, O, S\, O_3) + aq = K\, O, S\, O_3 + Mn\, O, S\, O_3 + aq + \mathbf{I}.$

211.
$$H\, O, K\, O, C_8\, H_4\, O_{10} \qquad\qquad = \text{Cream of Tartar.}$$
$$Na\, O, K\, O, C_8\, H_4\, O_{10} \qquad\quad = \text{Rochelle Salts.}$$
$$[N\, H_4]\, O, K\, O, C_8\, H_4\, O_{10} = \text{Ammoniated Tartar.}$$
$$Fe_2\, O_3, K\, O, C_8\, H_4\, O_{10} \qquad = \text{Tartarized Iron}$$
$$Sb_2\, O_3, K\, O, C_8\, H_4\, O_{10} \qquad = \text{Tartar Emetic.}$$
$$B\, O_3\, K\, O, C_8\, H_4\, O_{10} \qquad\quad = \text{Soluble Cream of Tartar.}$$

213. $4\, (K\, O, C\, O_2) + 16\, S = K\, O, S\, O_3 + 3\, K\, S_5 + 4\, \mathbf{C\, O_2}.$

$\qquad 3\, (K\, O, C\, O_2) + 8\, S = K\, O, S_2\, O_2 + 2\, K\, S_3 + 3\, \mathbf{C\, O_2}.$

The first or the last of these reactions takes place, according to the proportion of sulphur and the degree of temperature to which the mixture is exposed. If the sulphur is in excess, and the temperature of the mass raised to a red heat, pentasulphuret of potassium and sulphate of potassa are formed. If, however, the sulphur is present in smaller quantity, and the temperature of the mass is not raised above its melting point, a mixture of tersulphide of potassium and hyposulphite of potassa results. At the higher temperature, hyposulphite of potassa, which is always first formed, splits up into sulphate of potassa and pentasulphide of potassium. Thus,

$$4\, (K\, O, S_2\, O_2) = 3\, K\, O, S\, O_3 + K\, S_5.$$

Either of these sulphides is decomposed by dilute acids, forming sulphide of hydrogen, and setting free as many equivalents of sulphur, less one, as are contained in the compound.

$$K\,S_n + H\,O,\ S\,O_3 + Aq = S_{n-1} + K\,O,\ S\,O_3 + Aq + \mathbf{H\,S}.$$

$$K\,O,\ S\,O_3 + 4\,C = K\,S + 4\,\mathbf{C\,O}.$$

Sodium (Na).

215. *Problems on Common Salt.*

1. One gramme of salt contains how much chlorine and how much sodium?

2. One cubic centimetre of common salt, Sp. Gr. = 2.13, contains how many cubic centimetres of sodium, and how many of chlorine gas?

218. $Na\,Cl + H\,O,\ S\,O_3 = Na\,O,\ S\,O_3 + \mathbf{H\,Cl}.$

219. $Na\,O,\ S\,O_3 + 4\,C = Na\,S + 4\,\mathbf{C\,O}.$

220. $Na\,S + Ca\,O,\ C\,O_2 = Ca\,S + Na\,O,\ C\,O_2.$

$$3\,(Na\,O,\ S\,O_3) + 13\,C + 4\,(Ca\,O,\ C\,O_2) = 3\,Na\,O,\ C\,O_2 + 3\,Ca\,S,\ Ca\,O + 14\,\mathbf{C\,O}.$$

1. How much carbonate of soda can be made from 500 kilogrammes of common salt? How much sulphuric acid? How much charcoal and how much chalk are required in the process, according to the theory?

2. If in a chemical process carbonate of potassa or carbonate of soda may be used indifferently, which would be employed most profitably if the price were the same?

3. What relation ought the price of crystallized carbonate of soda to bear to that of the dry salt, if the intrinsic value is alone considered?

221. $Na\,O,\ C\,O_2 + Ca\,O,\ H\,O + Aq = Ca\,O,\ C\,O_2 + Na\,O,\ H\,O_2 + Aq.$

222. $Na\,O,\,C\,O_2 + 2\,C = Na + 3\,\mathbf{C}\,\mathbf{O}.$

223. $2\,(Na\,O,\,C\,O_2) + 3\,H\,O,\,P\,O_5 + Aq =$
$\quad H\,O,\,2\,Na\,O,\,P\,O_5 + Aq + 2\,\mathbf{C}\,\mathbf{O}_2.$

The symbol of crystallized phosphate of soda $=$
$H\,O,\,2\,Na\,O,\,P\,O_5\,.\,24\,H\,O.$

When heated.

$H\,O,\,2\,Na\,O,\,P\,O_5\,.\,24\,H\,O = 2\,Na\,O,\,P\,O_5 + 25\,\mathbf{H}\,\mathbf{O}.$

$3\,(Ag\,O,\,N\,O_5) + H\,O,\,2\,Na\,O,\,P\,O_5 + Aq =$
Yellow.
$3\,Ag\,O,\,P\,O_5 + 2\,(Na\,O,\,N\,O_5) + H\,O,\,N\,O_5 + Aq.$

White.
$2\,(Ag\,O,\,N\,O_5) + 2\,Na\,O,\,P\,O_5 + Aq = 2\,Ag\,O,\,P\,O_5$
$+ 2\,(Na\,O,\,N\,O_5) + Aq.$

1. How much sodium can be made from one kilogramme of anhydrous carbonate of soda?

2. How much $Ag\,O,\,N\,O_5$ is required to precipitate completely the phosphoric acid from 1.345 grammes of crystallized phosphate of soda? How much, from the same amount of pyrophosphate of soda?

224. $Na\,O,\,N\,O_5 + 2\,(H\,O,\,S\,O_3) = Na\,O,\,S\,O_3,\,H\,O,\,S\,O_3$
$+ \mathbf{H}\,\mathbf{O},\,\mathbf{N}\,\mathbf{O}_5.$

$Na\,O,\,N\,O_5 + K\,Cl + aq = Na\,Cl + K\,O,\,N\,O_5 + aq.$

225. $Na\,O,\,2\,B\,O_3\,.\,10\,H\,O =$ Symbol of common borax.

$Na\,O,\,2\,B\,O_3\,.\,5\,H\,O =$　"　　"　octohedral borax.

1. When soda nitre and potash nitre bear both the same price, from which can nitric acid be most profitably extracted?

2. When potash nitre is worth 25 cents the kilogramme, and chloride of potassium 8 cents the kilogramme, at what

price of soda nitre will it be profitable to convert it into potash nitre, assuming that the cost of the process amounts to two cents on each kilogramme of potash nitre manufactured?

3. What is the percentage composition of common borax? What is that of octohedral borax?

4. How much borax glass can be made from 100 grammes of common borax?

Ammonia ($[N H_4] O$).

The above is the symbol of the base of the ammonia salts, and corresponds to $K O$ and $Na O$. The student must be careful not to confound it with *ammonia gas*, which has the symbol $N H_3$.

227. $K O, H O + Fe = K O + Fe O + \mathbf{H}$.

$K O, N O_5 + 5 Fe = K O + 5 Fe O + \mathbf{N}$.

$3 (K O, H O) + K O, N O_5 + 8 Fe = 4 K O + 8 Fe O + \mathbf{N H_3}$.

229. $\mathbf{N H_3} + \mathbf{H Cl} = [N H_4] Cl$.

$[N H_4] Cl + Ca O, H O = Ca Cl + 2 H O + \mathbf{N H_3}$.

230. $Aq + \mathbf{N H_3} = [N H_4] O, H O + Aq$.

231. $[N H_4] O, H O + Aq + 2 \mathbf{H S} = [N H_4] S, H S + Aq$.

The compound which remains in solution after passing $H S$ gas through aqua ammonia until saturation, is $[N H_4] S, H S$, the sulphohydrate of sulphide of ammonium, which is the reagent so much used in the laboratory, and incorrectly called sulphide of ammonium. On exposure to the air, the solution becomes yellow, owing to the formation of bisulphide of ammonium, as is shown in the reaction below.

$$[NH_4]\,S, H\,S + [NH_4]\,O, HO' + Aq =$$
$$2\,[NH_4]\,S + Aq.$$

Colorless Solution.
$$2\,([NH_4]\,S,\,H\,S) + Aq + 5\,\mathbf{O} =$$

Yellow Solution.
$$[NH_4]\,S_2 + [NH_4]\,O, S_2\,O_2 + Aq.$$

232. $2\,[NH_4]\,Cl + 3\,(Ca\,O, C\,O_2) = 3\,Ca\,Cl + \mathbf{NH_3}$
$+ 2\,\mathbf{[NH_4]\,O}, 3\,\mathbf{C\,O_2} + \mathbf{H\,O}.$

1. How many cubic centimetres do 17 grammes of ammonia gas occupy? How many do 36.5 grammes of chlorohydric acid gas occupy? When the two gases combine, in what proportions, by volume, do they unite? How great is the condensation which results? Sp. Gr. of $[N\,H_4]\,Cl = 1.5$.

2. How much ammonia gas can be obtained from 5 grammes of chloride of ammonium. How much chloride of ammonium and how much $Ca\,O$ must be used in order to prepare one litre of ammonia gas?

3. How much chloride of ammonium and how much lime must be used in order to prepare one litre of aqua ammonia of Sp. Gr. $= 0.9$? How much water must be placed in the receiver?

4. How much ammonia gas is held in solution by one litre of aqua ammonia of Sp. Gr. 0.9? To how much $[N\,H_4]\,O, H\,O$ does this amount of gas correspond?

5. How much $H\,O, S\,O_3$ is required to neutralize four cubic centimetres of aqua ammonia of Sp. Gr. 0.9?

6. How much $H\,S$ gas will be absorbed by one litre of aqua ammonia of Sp. Gr. 0.9, assuming that only so much is taken up as is necessary to form the compound

[N H$_4$] S, H S? How much H O, S O$_3$ and Fe S will be required to produce this amount of gas? How much additional aqua ammonia must be added to the above solution in order to change it to a solution of proto-sulphide of ammonium?

SECOND GROUP: THE ALKALINE EARTHS.

Calcium (Ca).

241. $Ca\, O, S\, O_3 + Ba\, Cl + Aq = \text{Ba O, S O}_3 + Ca\, Cl + Aq.$

$Ca\, O,\ S\, O_3\ +\ H\, O,\ C_2\, O_3\ +\ Aq\ =\ \text{Ca O, C}_2\, O_3 + H\, O, S\, O_3 + Aq.$

242. $H\, O, 2\, Na\, O, P\, O_5 + 2\, Ca\, Cl + Aq = H\, O, 2\, \text{Ca O}, P\, O_5 + 2\, Na\, Cl + Aq.$

244. $2\,(\text{Ca O, H O}) + 2\,\mathbf{Cl} = \text{Ca O, Cl O} + \text{Ca Cl.}$

246. $\text{Ca O, C O}_2 + H\, Cl + Aq = Ca\, Cl + Aq + \mathbf{C\, O_2}.$

1. How much Ca O can be obtained from 100 kilogrammes of carbonate of lime? How much Ca O, H O can be obtained from the same amount?

2. How much carbonate of lime must be burnt in order to yield 140 kilogrammes of quicklime? How much to yield 185 kilogrammes of Ca O, H O? How many cubic metres of C O$_2$ would be set free during the process?

3. How many cubic metres of C O$_2$ can be absorbed by a quantity of milk of lime containing 5 kilogrammes of Ca O?

4. What is the percentage composition of unburnt and of burnt gypsum?

5. What is the percentage composition of phosphate of lime ($3\,Ca\,O,\,P\,O_5$)?

6. How much chlorine and how much lime are required to make 100 kilogrammes of chloride of lime, assuming that its composition is expressed by the symbol $Ca\,O,\,Cl\,O + Ca\,Cl$? How much $Mn\,O_2$ and how much muriatic acid of Sp. Gr. 1.15 will yield the requisite amount of chlorine?

7. To how much $Ca\,O$, how much $Ca\,Cl$, and how much $Ca\,O,\,S\,O_3$ do 2.5 grammes of carbonate of lime correspond?

Barium and Strontium (Ba and Sr).

248. $Ba\,O,\,S\,O_3 + 4\,C = Ba\,S + 4\,\mathbf{C\,O}.$

$Ba\,S + H\,Cl + Aq = Ba\,Cl + Aq + \mathbf{H\,S}.$

$Cu\,O + Ba\,S + Aq = Cu\,S + Ba\,O,\,H\,O + Aq.$

$Ba\,Cl + Na\,O,\,S\,O_3 + Aq = Ba\,O,\,S\,O_3 + Na\,Cl + Aq.$

$Sr\,O,\,S\,O_3 + 4\,C = Sr\,S + 4\,\mathbf{C\,O}.$

$Sr\,S + H\,O,\,N\,O_5 + Aq = Sr\,O,\,N\,O_5 + Aq + \mathbf{H\,S}.$

$Sr\,O,\,N\,O_5 + Na\,O,\,S\,O_3 + Aq = Sr\,O,\,S\,O_3\;Na\,O,\,N\,O_5 + Aq.$

1. What is the percentage composition of sulphate of baryta?

2. How much chloride of barium can be made from 5 kilogrammes of sulphate of baryta? How much nitrate of baryta, and how much $Ba\,O$, can be made from the same amount of sulphate of baryta?

3. To how much sulphuric acid do 4.567 grammes of sulphate of baryta correspond? To how much chloride of barium do they correspond?

4. How much nitrate of strontia can be made from one kilogramme of sulphate of strontia? How much carbon and how much nitric acid of Sp. Gr. 1.4 is required in the process?

Magnesium (Mg).

249.   The symbol of serpentine mineral is

$$9\,[\text{Mg, Fe}]\,\text{O},\,4\,\text{Si}\,\text{O}_3,\,6\,\text{H}\,\text{O}.$$

The symbol [Mg, Fe] O indicates that a portion of the magnesium in the base is replaced by oxide of iron, and the whole stands for but one equivalent of base. We frequently write, instead of the portion enclosed in brackets, the general symbol R, when the symbol of the mixed base becomes R O, and that of serpentine mineral $9\,\text{R}\,\text{O},\,4\,\text{Si}\,\text{O}_3$. It must be noticed, that Mg and Fe when enclosed in brackets, with a comma between, as above, no longer stand for an equivalent of each metal. On the other hand, the two together make but one equivalent of metal, represented by R in the other mode of writing. Moreover, nothing is intended to be indicated in regard to the relative proportions of the two metals mixed together to form one equivalent, as they vary in different specimens of the same mineral. This method of writing the symbols of compounds containing isomorphous constituents will be constantly used hereafter.

$$9\,[\text{Mg, Fe}]\,\text{O},\,4\,\text{Si}\,\text{O}_3\,.\,6\,\text{H}\,\text{O} + 9\,(H\,O,\,S\,O_3) + Aq =$$
$$4\,\text{Si}\,\text{O}_3 + 9\,[Mg,\,Fe]\,O,\,S\,O_3 + Aq.$$

Since sulphate of magnesia and sulphate of protoxide of iron have the same crystalline form, we shall obtain, on evaporating the solution, crystals containing both salts. We can prevent the sulphate of protoxide of iron from crystallizing, by converting it into sulphate of sesquioxide of iron by means of nitric acid. Thus,

$$6 (Fe\,O,\,S\,O_3) + 3\,(H\,O,\,S\,O_3) + H\,O,\,N\,O_5 + Aq =$$
$$3\,(Fe_2\,O_3,\,3\,S\,O_3) + Aq + \mathbf{N\,O_2}.$$

Talc $\qquad = 6\,Mg\,O,\,5\,Si\,O_3,\,2\,H\,O.$

Meerschaum $= Mg\,O,\,Si\,O_3,\,H\,O.$

Hornblende $= 4\,[Mg,\,Ca,\,Fe]\,O,\,3\,Si\,O_3.$

Augite $\qquad = 3\,[Ca,\,Fe,\,Mg]\,O,\,2\,Si\,O_3.$

250. $5\,(Mg\,O,\,S\,O_3) + 5\,(KO,\,CO_2) + Aq = 3\,(MgO,\,CO_2.aq)$
$$+ Mg\,O,\,H\,O + Mg\,O,\,2\,CO_2 + 5\,(KO,\,S\,O_3) + Aq.$$

The relative proportions of carbonate of magnesia and of hydrate of magnesia vary with the temperature and other circumstances attending the precipitation.

251. $Mg\,O,\,CO_2 + HCl + Aq = Mg\,Cl + Aq + \mathbf{CO_2}.$

$$2\,(Mg\,O,\,S\,O_3) + H\,O,\,2\,Na\,O,\,P\,O_5 + [N\,H_4]\,O,\,H\,O$$
$$+ Aq = [N\,H_4]\,O,\,2\,Mg\,O,\,P\,O_5 + 2\,(Na\,O,\,S\,O_3)$$
$$+ Aq.$$

When heated.
$$(N\,H_4)\,O,\,2\,Mg\,O,\,P\,O_5 = 2\,Mg\,O,\,P\,O_5 + \mathbf{N\,H_3} + \mathbf{H\,O}.$$

1. What is the percentage composition of talc? What that of hornblende and augite, assuming that the whole of the base in either case is $Mg\,O$?

2. From a solution of sulphate of magnesia the whole of the magnesia was precipitated by phosphate of soda and ammonia. This precipitate, after ignition, was found to weigh 2.456 grammes. How much sulphate of magnesia was contained in the solution?

Aluminum (Ál).

258. $Al_2 O_3, Si O_3 . 2 H O + 3 (H O, S O_3) + aq = Si O_3 + Al_2 O_3, 3 S O_3 + aq.$

$Al_2 O_3, 3 S O_3 + 3 (Na O, C O_2) + Aq = Al_2 O_3, 3 H O + 3 (Na O, S O_3) + Aq + \mathbf{C O_2}.$

$Al_2 O_3, 3 H O + K O, H O + Aq = K O, Al_2 O_3 + Aq.$

$K O, S O_3 . Al_2 O_3, 3 S O_3 + 3 (Na O, C O_2) + Aq = Al_2 O_3, 3 H O + 3 (Na O S O_3) + K O, S O_3 + Aq + \mathbf{C O_2}.$

$Al_2 O_3, 3 S O_3 + 3 (Pb O, [C_4 H_3] O_3) + Aq = 3 (Pb O, S O_3) + Al_2 O_3, 3 [C_4 H_3] O_3 + Aq.$

Symbols of Isomorphous Alums.

Potassa, Alumina, Alum.
$K O, S O_3 . Al_2 O_3, 3 S O_3 . 24 H O.$

Soda, Alumina, Alum.
$Na O, S O_3 . Al_2 O_3, 3 S O_3 . 24 H O.$

Ammonia, Alumina, Alum.
$[N H_4] O, S O_3 . Al_2 O_3, 3 S O_3 . 24 H O.$

Potassa, Chrome, Alum.
$K O, S O_3 . Cr_2 O_3, 3 S O_3 . 24 H O.$

Soda, Chrome, Alum.
$Na O, S O_3 . Cr_2 O_3, 3 S O_3 . 24 H O.$

Ammonia, Chrome, Alum.
$[N H_4] O, S O_3 . Cr_2 O_3, 3 S O_3 . 24 H O.$
8 *

Potassa, Iron, Alum.

$$K O, S O_3 . Fe_2 O_3, 3 S O_3 . 24 H O.$$

Soda, Iron, Alum.

$$Na O, S O_3 . Fe_2 O_3, 3 S O_3 . 24 H O.$$

Ammonia, Iron, Alum.

$$[N H_4] O, S O_3 . Fe_2 O_3, 3 S O_3 . 24 H O.$$

Symbols of the most important Silicates of Alumina.

$2 Al_2 O_3, Si O_3,$	Staurotide.
$3 Al_2 O_3, 2 Si O_3,$	Andalusite.
$3 Al_2 O_3, 2 Si O_3,$	Kyanite.
$3 Al_2 O_3, 2 Si [O, Fl]_3,$	Topaz.
$K O, Si O_3 . Al_2 O_3, 3 Si O_3,$	Common Felspar.
$Na O, Si O_3 . Al_2 O_3, 3 Si O_3,$	Albite.
$[Ca, Na] O, Si O_3 . Al_2 O_3, Si O_3,$	Labradorite.
$K O, Si O_3 . 4 (Al_2 O_3, Si O_3),$	Common Mica.
$3 R O,* Si O_3 . R_2 O_3,† Si O_3,$	Garnet.

1. What is the percentage composition of staurotide? What is that of kyanite?

2. An analysis of one of the above silicates would give the following percentage composition.

Silica,	64.76
Potassa,	16.87
Alumina,	18.37
	100.00

What is the symbol of the mineral?

---

* R O = Fe O, Mn O, Mg O, or Ca O.
† $R_2 O_3$ = $Al_2 O_3$, Fe_2, O_3, or $Cr_2 O_3$.

Solution. — This problem is evidently the reverse of deducing the per_centage composition from the symbol; but it does not admit, like that, of a definite solution, for while there is but one percentage composition corresponding to a given symbol, there may be an infinite number of symbols corresponding to a given percentage composition. This can easily be made clear by an example. The commonly received symbol of alcohol is [$C_4 H_5$] O, $H O = C_4 H_6 O_2$. The percentage composition is easily ascertained. Thus,

$$C_4 \quad H_6 \quad O_2$$
$$24 + 6 + 16 = 46.$$

$46 : 24 = 100 : x = 52.18$ per cent of carbon.
$46 : 6 = 100 : x = 13.04$ per cent of hydrogen.
$46 : 16 = 100 : x = 34.78$ per cent of oxygen.

	Per cent.					
Carbon,	$52.18 = C_2 = 12$	or	$C_4 = 24$	or	$C_6 = 36$
Hydrogen,	$13.04 = H_3 = 3$	"	$H_6 = 6$	"	$H_9 = 9$
Oxygen,	$34.78 = O = 8$	"	$O_2 = 16$	"	$O_3 = 24$
	100.00		23		46	69

This percentage composition evidently corresponds not only to $C_4 H_6 O_2$, but also to $C_2 H_3 O$, to $C_6 H_9 O_3$, and to any other symbol which is a multiple of the first; for, taking the per cent of carbon as an example, we have

$$100 : 52.18 = 23 : 12 = 46 : 24 = 69 : 36 = 92 : 48, \&c.$$

If, then, we had given the percentage composition of alcohol, it would be impossible to determine, without other data, whether the symbol was $C_2 H_3 O$, or some multiple of it. If, however, we had also given that the sum of the equivalents of the elements of alcohol equalled 46, then we could easily reverse the above process. Thus,

$100 : 52.18 = 46 : x = 24$, the sum of the equivalents of carbon.
$100 : 13.04 = 46 : x = 6$,   "   "   "   "   hydrogen.
$100 : 34.78 = 46 : x = 16$,   "   "   "   "   oxygen.

And $\quad \dfrac{24}{6} = 4$, number of equivalents of carbon.

$\dfrac{6}{1} = 6$,   "   "   "   "   hydrogen.

$\dfrac{16}{8} = 2$,   "   "   "   "   oxygen.

Assuming, however, that we had no means of ascertaining the sum of the equivalents in alcohol, then, although we could not definitely fix its symbol, yet nevertheless we could easily find which of all the possible symbols expressed its composition in the simplest terms; in other words, with the fewest number of whole equivalents. For this purpose, assume for a moment that the sum of the equivalents is equal to 100, then

52.18 = the sum of the equivalents of carbon.

13.04 =   "   "   "   "   hydrogen.

34.78 =   "   "   "   "   oxygen.

$$\frac{52.18}{6} = 8.697, \text{ number of equivalents of carbon.}$$

$$\frac{13.04}{1} = 13.04, \quad " \quad " \quad " \quad " \text{ hydrogen.}$$

$$\frac{34.78}{8} = 4.348, \quad " \quad " \quad " \quad " \text{ oxygen.}$$

These are the number of equivalents of each element on the supposition that the sum of the equivalents in alcohol is equal to 100. Any other possible number of equivalents must be either a multiple or a submultiple of these, and we can easily find the fewest number of whole equivalents possible, by seeking for the three smallest whole numbers which stand to each other in the relation of   8.697 : 13.08 : 4.348, which will be found to be   2 : 3 : 1.
Hence the simplest possible symbol is $C_2 H_3 O$, but, from anything we are assumed to know, the symbol may be any multiple of this; and for considerations which cannot be discussed in this connection, chemists usually assign to alcohol the symbol $C_4 H_6 O_2$, which is double the above.

The symbol thus obtained expresses merely the relative number of equivalents of each element present in the compound, and gives no information in regard to the grouping of the elements. Such symbols are called empirical symbols, to distinguish them from the rational symbols, which indicate the manner in which the elements are supposed to be arranged. The rational symbol of alcohol is $[C_4 H_5] O, H O$. This indicates not only that alcohol consists of four equivalents of carbon, six of hydrogen, and two of oxygen, but also that it is the hydrated oxide of a compound radical called ethyle. It must be carefully noticed, however, that the empirical symbols fully represent all our positive knowledge. They alone are not liable to be changed. The grouping of elements in a compound is a matter of theory, and the rational symbols are liable to constant changes, as the opinions of chemists on this subject vary.

From the example just discussed we can easily deduce the following rule for finding the empirical symbol of a compound from its percentage composition. *Divide the per cent of each element entering into the compound by its chemical equivalent, and find the simplest series of whole numbers to which these results correspond.* To apply this rule to the problem under consideration.

$$\frac{64.76}{45.3} = 1.43, \text{ number of equivalents of silica.}$$

$$\frac{16.87}{47.2} = 0.3575, \quad " \quad " \quad " \quad " \quad \text{potassa.}$$

$$\frac{18.37}{51.4} = 0.3575, \quad " \quad " \quad " \quad " \quad \text{alumina.}$$

$$143 : 0.3575 : 0.3575 = 4 : 1 : 1.$$

Empirical symbol, $Al_2 O_3$, $K O$, $4 Si O_3$.

Rational symbol, $K O$, $Si O_3$, $Al_2 O_3$, $3 Si O_3$.

3. An analysis of one of the silicates of alumina would give the following percentage composition.

	Per cent.
Silica,	53.29
Lime,	16.47
Alumina,	30.24
	100.00

What is the symbol of the mineral?

Ans. $Ca O$, $Si O_3$. $Al_2 O_3$ $Si O_3$.

Second Method of Solution. — By inspecting the formula obtained by solving the problem according to the method just described, the student will see that the amount of oxygen in the acid stands in a very simple relation to that in the bases. This relation is $1 : 3 : 6$, corresponding to $Ca O$, $Al_2 O_3$, and $2 Si O_3$. It has been shown in the text-book, § 200, that a similar simple ratio exists between the amount of oxygen in the acid and that in the bases of all oxygen salts. The ratio can easily be found from the percentage composition. For this purpose we have merely to calculate the amount of oxygen in the per cent of the acid and bases indicated by analysis, and find the simplest ratio in which these amounts stand to each other. In our example,

53.29 per cent of silica contains 28.24 parts of oxygen.
16.47 " " lime " 4.71 " " "
30.24 " " alumina " 14.12 " " "

According to the principle just stated, these numbers ought to stand to each other in some simple ratio, and it can easily be seen that

$$4.71 : 14.12 : 28.24 = 1 : 3 : 6.$$

From this ratio we can easily deduce the symbol, for one equivalent of oxygen corresponds to one equivalent of $Ca O$, three equivalents of oxygen correspond to one equivalent of $Al_2 O_3$, and six of oxygen to two equivalents of $Si O_3$. Hence the symbol is $Ca O, Al_2 O_3, 2 Si O_3$, which we may write' as above, $Ca O, Si O_3 . Al_2 O_3, Si O_3$. For convenience in calculating the amount of oxygen from the per cent of acids or bases indicated by analysis, Table VII. has been added at the end of the book, which gives the per cent of oxygen, together with its logarithm contained in the bases and acids of most common occurrence.

In deducing empirical symbols from the results of actual analysis, it must be remembered that our processes are not absolutely accurate, and that therefore we must not expect to find more than a close approximation to a simple ratio between the oxygen in the base and that in the acid. Again, in mineral compounds, it is very frequently the case that isomorphous bases replace each other to a greater or less extent. This is the case in common garnet, the symbol of which may be written thus:

$$3 [Fe, Mn, Mg, Ca] O, Si O_3 . [Al_2 Fe_2] O_3, Si O_3.$$

We generally, however, write the symbol as on page 90:

$$3 R O, Si O_3 . R_2 O_3, Si O_3.$$

Here $R O$ stands for the sum of all the protoxide bases, which make together but one equivalent of base, and $R_2 O_3$ for the sum of all the sesquioxide bases, which also make together but one equivalent of base. Such general symbols as these give all the information in regard to the constitution of the mineral which is required. In deducing such symbols, it is evident that the oxygen of all the protoxide bases must be added together to obtain the amount of oxygen in the assumed base $R O$, and all the oxygen of the sesquioxide must be added together in the same way in order to obtain the amount of oxygen in the assumed base $R_2 O_3$. From these sums we can easily obtain the required oxygen ratio.

4. An analysis of andesine (a mineral allied to felspar) yielded the following result.

		Proportion of Oxygen.	
Silicic Acid,	59.60		30.90 in $Si O_3$.
Alumina,	24.28	11.22	$= 11.70$ in $R_2 O_3$.
Sesquioxide of Iron,	1.58	0.48	
Lime,	5.77	1.61	
Magnesia,	1.08	0.37	$= 3.79$ in $R O$.
Soda,	6.53	1.65	
Potassa,	1.08	0.16	
	99.92		

What is the symbol of the mineral?

Solution. — The ratio of the oxygen in R O, R$_2$ O$_3$, and Si O$_3$ is

$$3.79 : 11.70 : 30.90 = 1 : 3.08 : 8.1,$$

for which we may substitute, for reasons stated above,

$$1 : 3 : 8.$$

One equivalent of oxygen corresponds to R O.

Three equivalents of oxygen correspond to R$_2$ O$_3$.

Eight equivalents of oxygen correspond to $\frac{8}{3}$ Si O$_3$.

But as we do not admit fractional equivalents, we may multiply the whole by three, when we obtain the empirical symbol

$$3 \text{ R O}, 3 \text{ R}_2 \text{ O}_3, 8 \text{ Si O}_3;$$

from which we may deduce the rational symbol

$$3 \text{ R O}, 2 \text{ Si O}_3 . 3 (\text{R}_2 \text{ O}_3, 2 \text{ Si O}_3).$$

5. Deduce the symbols of the silicious minerals of which the following are analyses.

	1.	2.	3.	4.
Silicic Acid,	65.72	68.4	44.12	63.70
Sesquioxide of Iron,		0.1	0.70	0.50
Alumina,	18.57	20.8	35.12	23.95
Lime,	0.34	0.2	19.02	2.05
Magnesia,	0.10		0.56	0.65
Potassa,	14.02		0.25	1.20
Soda,	1.25	10.5	0.27	8.11
	100.00	100.0	100.04	100.16

HEAVY METALS.

FIRST GROUP OF THE HEAVY METALS.

Iron (Fe).

286. *a.* $4 \text{ Fe} + \mathbf{O} = \text{Fe}_4 \text{ O}.$

b. $3 \text{ Fe}_4 \text{ O} + 13 \mathbf{O} = 4 \text{ (Fe O, Fe}_2 \text{ O}_3).$

c. $2 \text{ (Fe O, Fe}_2 \text{ O}_3) + \mathbf{O} = 3 \text{ Fe}_2 \text{ O}_3.$

d. $2 \text{ (Fe O, S O}_3 . \text{ 6 H O)}$ when heated resolves into $F_2 O_3, \text{ S O}_3 + \mathbf{S\,O_2} + 12 \,\mathbf{H\,O}$; by further heating, $\text{Fe}_2 \text{ O}_3, \text{ S O}_3 = \text{Fe}_2 \text{ O}_3 + \mathbf{S\,O_3}.$

e. $3 \text{ Fe} + 4 \mathbf{O} = \overset{\text{Red.}}{\text{Fe O}}, \text{Fe}_2 \text{ O}_3.$

f. $2 \text{ (Fe O, Fe}_2 \text{ O}_3) + 9 HO + \mathbf{O} = 3 \,(\overset{\text{Red.}}{\text{Fe}_2 O_3}, 3 \text{ H O)}.$

g. $\text{Fe O, Fe}_2 \text{ O}_3 + 2 \text{ } C \text{ } O_2 + Aq = \text{Fe}_2 \text{ O}_3 + Fe \text{ O}, 2 \text{ } C \text{ } O_2 + Aq.$

$2 \text{ } (Fe \text{ O}, 2 \text{ } C \text{ } O_2) + Aq + \mathbf{O} = \overset{\text{Red.}}{\text{Fe}_2 O_3}, 3 \text{ H O} + 4 \text{ } C \text{ } O_2 + Aq.$

285. $\text{Fe} + HO, S \text{ } O_3 + Aq = Fe \text{ O}, S \text{ } O_3 + Aq + \mathbf{H}.$

285. $\quad Fe + Cu\ O,\ S\ O_3 + Aq = Cu + Fe\ O,\ S\ O_3$

$\quad\quad + Aq.$

$\quad$ a. $\ 6\ (Fe\ O,\ S\ O_3) + Aq + 3\ \mathbf{O} = Fe_2\ O_3,\overset{Red.}{}\ 3\ H\ O$

$\quad\quad + 2\ (Fe_2\ \overset{Yellow\text{-}Brown.}{O_3,\ 3\ S\ O_3}) + Aq.$

$\quad$ b. $\ 6\ (Fe\ O,\ S\ O_3) + 3\ (H\ O,\ S\ O_3) + H\ O,\ N\ O_5$

$\quad\quad + Aq = 3\ (Fe_2\ \overset{Yellow\text{-}Brown.}{O_3,\ 3\ S\ O_3}) + Aq + \mathbf{N\ O_2}.$

$\quad$ c. $\ Fe\ O,\ S\ O_3 + [N\ H_4]\ O,\ H\ O + Aq = \overset{Light\ Green.}{Fe\ O,\ H\ O}$

$\quad\quad + [N\ H_4]\ O,\ S\ O_3 + Aq.$

$\quad Fe_2\ O_3,\ 3\ S\ O_3 + 3\ ([N\ H_4]\ O,\ H\ O) + Aq =$

$\quad Fe_2\ O_3,\overset{Red.}{}\ 3\ H\ O + 3\ ([N\ H_4]\ O,\ S\ O_3) + Aq.$

$\quad$ d. $\ 2\ Fe + 4\ H\ O,\ N\ O_5 + Aq = Fe_2\ O_3,\ 3\ N\ O_5 + Aq$

$\quad\quad + \mathbf{N\ O_2}.$

288. $\quad [H\ O,\ 2\ Na\ O]\ P\ O_5 + 3\ (Fe\ O,\ S\ O_3) + Aq =$

$\quad\quad 3\ \overset{White.}{Fe\ O,}\ P\ O_5 + 2\ Na\ O,\ S\ O_3 + H\ O,\ S\ O_3$

$\quad\quad + Aq.$

$\quad [H\ O,\ 2\ Na\ O]\ P\ O_5 + Fe_2\ O_3,\ 3\ S\ O_3 + Aq =$

$\quad \overset{White.}{Fe_2 O_3, P O_5.4HO} + 2(Na\ O, S O_3) + HO, SO_3 + Aq.$

290. $\quad 3\ Fe\ O + 2\ Fe_2\ O_3 + 9\ H\ Cy + Aq =$

$\quad\quad 3\ \overset{Blue.}{Fe\ Cy,}\ 2\ Fe_2\ Cy_3 + 9\ H\ O + Aq.$

291. $\quad 3\ \overset{Blue.}{Fe\ Cy,}\ 2\ Fe_2\ Cy_3 + 6\ (K\ O,\ H\ O) + Aq =$

$\quad\quad 2\ (Fe_2\ O_3,\overset{Red.}{}\ 3\ H\ O) + 3\ (2\ K\ Cy,\ Fe\ Cy) + Aq.$

292. a. $\ 3\ (2\ K\ Cy,\ Fe\ Cy) + 2\ (Fe_2\ O_3,\ 3\ S\ O_3) + Aq =$

$\quad\quad 3\ \overset{Blue.}{Fe\ Cy,}\ 2\ Fe_2\ Cy_3 + 6\ (K\ O,\ S\ O_3) + Aq.$

9

292. *b.* $2 K Cy, Fe Cy + 2 (Fe O, S O_3) + Aq =$

       White.

     $3 Fe Cy + 2 K O, \acute{S} O_3 + Aq.$

     White.   Blue.

$9 Fe Cy + 3 \; \mathbf{0} = Fe_2 O_3 + 3 Fe Cy, 2 Fe_2 Cy_3.$

c. $2 K Cy, Fe Cy + 2 (Cu O, S O_3) + Aq =$

       Purple.

     $2 Cu Cy, Fe Cy. Aq + 2 (K O, S O_3) + Aq.$

$2 K Cy, Fe Cy + 2 (Pb O, N O_5) + Aq =$

               White.

     $2 Pb Cy, Fe Cy. Aq + 2 (K O, N O_5) + Aq.$

A large number of similar compounds having the general symbol 2 R Cy, Fe Cy. Aq, may be formed thus : —

     White.   Pale Yellow.

2 H Cy, Fe Cy. Aq.   2 [N H_4] Cy, Fe Cy. Aq.

    Yellow.   Yellow.

2 Na Cy, Fe Cy. Aq.   2 Ba Cy, Fe Cy. Aq.

   Pale Yellow.   White.

2 Mg Cy, Fe Cy. Aq.   2 Zn Cy, Fe Cy. Aq.

There are also compounds in which the two equivalents of R in the general symbol are replaced by different metals thus : —

      White.   Yellow.

$\left\{ \begin{array}{l} Fe \; Cy \\ K \; Cy \end{array} \right\} Fe \; Cy.$ $\left\{ \begin{array}{l} Ba \; Cy \\ K \; Cy \end{array} \right\} Fe \; Cy. \; Aq.$

      Red.

$\left\{ \begin{array}{l} Cu \; Cy \\ K \; Cy \end{array} \right\} Fe \; Cy.$ $\left\{ \begin{array}{l} Ca \; Cy \\ K \; Cy \end{array} \right\} Fe \; Cy. \; Aq.$

     Yellow Salt.   Red Salt.

293. $2 (2 K Cy, Fe Cy) + Cl + Aq = 3 K Cy, Fe_2 Cy_3$

     $+ K Cl + Aq.$

$3 K Cy, Fe_2 Cy_3 + 3 (Fe O, S O_3) + Aq =$

       Blue.

$3 Fe Cy, Fe_2 Cy_3 + 3 (K O, S O_3) + Aq.$

293. There also may be formed a large number of similar compounds having the general symbol

$3 R Cy, Fe_2 Cy_3$, such as :—

$3 H \overset{\text{Brown.}}{Cy}, Fe_2 Cy_3.$ $3 Ca \overset{\text{Light-Red.}}{Cy}, Fe_2 Cy_3.$

$\left\{ \begin{array}{l} K\ Cy \\ 2\ Ba\ Cy \end{array} \right\}, \overset{\text{Red.}}{Fe_2\ Cy_3}.\ Aq.$

294. $\overset{\text{Light-Green Solution.}}{Fe\ O,\ S\ O_3} + [N\ H_4]\ S + Aq = \overset{\text{Black.}}{Fe\ S}$

   $+ [N\ H_4]\ O, S\ O_3 + Aq.$

295. $\overset{\text{Black.}}{Fe\ S} + aq + 4\ \mathbf{O} = \overset{\text{Light-Green.}}{Fe\ O,\ S\ O_3} + aq.$

Manganese (Mn).

299. $Mn\ O_2 + H\ O,\ S\ O_3 + aq = Mn\ O,\ S\ O_3 + aq$

   $+ \mathbf{O}.$

$\overset{\text{Black.}}{Mn\ O_2} + 2\ H\ Cl + Aq = \overset{\text{Light-Pink Solution.}}{Mn\ Cl} + Aq + \mathbf{Cl}.$

300. $a.$ $6\ (\overset{\text{Light-Pink Solution.}}{Mn\ O,\ S\ O_3}) + Aq + 3\ \mathbf{O} = \overset{\text{Brown.}}{Mn_2\ O_3, 3\ H\ O}$

   $+ 2\ (Mn_2\ O_3, 3\ S\ O_3) + Aq.$

$b.$ $\overset{\text{Light-Pink Solution.}}{Mn\ O,\ S\ O_3} + [N H_4]\ O,\ H\ O + Aq = \overset{\text{White.}}{Mn\ O, H\ O}$

   $+ [N\ H_4]\ O,\ S\ O_3 + Aq.$

$2\ (Mn\ \overset{\text{White.}}{O},\ H\ O) + aq + \mathbf{O} = Mn_2\ \overset{\text{Brown.}}{O_3, 3}\ H\ O$

   $+ aq.$

$\overset{\text{Light-Pink Solution.}}{Mn\ O,\ S\ O_3} + [N\ H_4]\ S + Aq = \overset{\text{Flesh-colored.]}}{Mn\ S}$

   $+ [N\ H_4]\ O,\ S\ O_3 + Aq.$

301. $Mn\ O_2 + \overset{\text{Melted together.}}{K\ O,\ H\ O} + \mathbf{O} = K\ O, Mn\ O_3 + \mathbf{H}\ \mathbf{O}.$

301. $3 \overset{\text{Green Solution.}}{(K\ O,\ Mn\ O_3)} + Aq + 2\mathbf{C\ O_2} = Mn\ O_2$

$+ \overset{\text{Crimson Solution.}}{K\ O,\ Mn_2\ O_7} + 2\ (K\ O,\ C\ O_2) + Aq.$

$3\ \overset{\text{Green Solution.}}{(K\ O,\ Mn\ O_3)} + 2\ (H\ O,\ S\ O_3) + Aq =$

$Mn\ O_2 + \overset{\text{Crimson Solution.}}{K\ O,\ Mn_2\ O_7} + 2\ (K\ O,\ S\ O_3) + Aq.$

Cobalt (Co) + Nickel (Ni).

303. $\overset{\text{Pink Solution.}}{Co\ O,\ S\ O_3} + [N\ H_4]\ S + Aq = \overset{\text{Black.}}{Co\ S}$

$[N\ H_4]\ O,\ S\ O_3 + Aq.$

$Ni\ \overset{\text{Green Solution.}}{O,\ S\ O_3} + [N\ H_4]\ S + Aq = \overset{\text{Black.}}{Ni\ S}$

$+ [N\ H_4]\ O,\ S\ O_3 + Aq.$

Zinc (Zn).

311. $Zn + H\ O,\ S\ O_3 + Aq = Zn\ O,\ S\ O_3 + Aq$

$+ \mathbf{H}.$

312. a. $Zn\ O,\ S\ O_3 + K\ O,\ H\ O + Aq = Zn\ \overset{\text{White.}}{O,}\ H\ O$

$+ K\ O,\ S\ O_3 + Aq.$

This precipitate dissolves in an excess of $K\ O,\ H\ O$

$+ Aq.$

b. $Zn\ O,\ S\ O_3 + [N\ H_4]\ S + Aq = \overset{\text{White.}}{Zn\ S}$

$+ [N\ H_4]\ O,\ S\ O_3 + Aq.$

312. *c.* $5 \, (Zn \, O, \, S \, O_3) + 5 \, (Na \, O, \, C \, O_2) + Aq =$

$2 \, (Zn \cdot \overset{\text{White.}}{O,} \, C \, O_2) + 3 \, (Zn \, \overset{\text{White.}}{O,} \, H \, O)$

$+ 5 \, (Na \, O, \, S \, O_3) + Aq + 3 \, \mathbf{C \, O_2}.$

This precipitate is a mixture of $Zn \, O$, $C \, O_2$, and $Zn \, O$, $H \, O$, but in variable proportions.

Cadmium (Cd).

315. $Cd \, O, \, S \, O_3 + H \, S + Aq = \overset{\text{Yellow.}}{Cd \, S} + H \, O, \, S \, O_3$

$+ \, Aq.$

Tin (Sn).

317. The oxides of tin are $\overset{\text{Dark-Brown.}}{Sn \, O}$; $\overset{\text{Gray-White.}}{Sn_2 \, O_3}$; and $\overset{\text{White.}}{Sn \, O_2}$.

319. $Sn + H \, Cl + aq = Sn \, Cl + aq + \mathbf{H}.$

320. $Sn \, Cl + [N \, H_4] \, O, \, H \, O + Aq = Sn \, \overset{\text{White.}}{O,} \, H \, O$

$+ [N \, H_4] \, Cl + Aq.$

321. $Sn \, Cl + Cl + Aq = Sn \, Cl_2 + Aq.$

$Sn + 2 \, H \, Cl + H \, O, \, N \, O_5 + Aq = Sn \, Cl_2$

$+ \, Aq + \mathbf{N \, O_3}.$

$Sn \, Cl_2 + 2 \, ([N \, H_4] \, O, \, H \, O) + Aq = H \, \overset{\text{White.}}{O,} \, Sn \, O_2$

$+ \, 2 \, [N \, H_4] \, Cl + Aq.$

$Sn \, O + K \, O, \, H \, O + Aq = K \, O, \, Sn \, O + Aq.$

$H \, O, \, Sn \, O_2 + K \, O, \, H \, O + Aq = K \, O, \, Sn \, O_2$

$+ \, Aq.$

9 *

By evaporation of the last solution we can obtain crystals of $K O, Sn O_2$. 4 H O.

We may also prepare $Na O, Sn O_2$. 4 H O.

324. $5 Sn + 10 (H O, N O_5) + Aq = \overset{\text{White.}}{Sn_5 O_{10}}$. 10 H O $+ Aq + 10 \mathbf{N O_4}$.

$Sn_5 O_{10}$. 10 H O $+ K O, H O + Aq = K O, Sn_5 O_{10} + Aq$.

By evaporation we can obtain crystals of $\overset{\text{White.}}{K O, Sn_5 O_{10}}$. 4 H O.

325. $Sn Cl + H S + Aq = \overset{\text{Brown.}}{Sn S} + H Cl + Aq$.

$Sn Cl_2 + 2 H S + Aq = \overset{\text{Yellow.}}{Sn S_2} + 2 H Cl + Aq$.

SECOND GROUP OF THE HEAVY METALS.

Lead (Pb).

331. The oxides of lead are $\overset{\text{Black.}}{Pb_2 O}$; $\overset{\text{Red or Yellow.}}{Pb O}$; $\overset{\text{Red-Yellow.}}{Pb_2 O_3}$; and $\overset{\text{Dark-Brown.}}{Pb O_2}$.

334. $3 Pb + 4 (H O, N O_5) + aq = 3 (Pb O, N O_5) + aq + \mathbf{N O_2}$.

$3 Pb O + 3 (H O, N O_5) + aq = 3 (Pb O, N O_5) + aq$.

335. $Pb\ O,\ N\ O_5 + H\ O,\ S\ O_3 + Aq = \overset{\text{White.}}{\text{Pb}}\ \text{O},\ \text{S}\ \text{O}_3$

$+ H\ O,\ N\ O_5 + Aq.$

$Al_2\ O_3,\ 3\ S\ O_3 + 3\ (Pb\ O,\ [C_4\ H_3]\ O_3) + Aq =$

$3\ (\overset{\text{White.}}{\text{Pb}}\ \text{O},\ \text{S}\ O_3) + Al_2\ O_3,\ 3\ [C_4\ H_3]\ O_3 + Aq.$

336. $\text{Pb O} + H\ Cl + aq = \text{Pb}\ \overset{\text{White.}}{\text{Cl}} + aq.$

$\text{Pb O} + H\ Cl + Aq = Pb\ Cl + Aq.$

337. Acetate of oxide of lead $= \text{Pb O},\ [C_4\ H_3]\ O_3 .\ 3\ \text{H O}.$

Basic acetate of oxide of lead $=$

$3\ \text{Pb O},\ [C_4\ H_3]\ O_3 .\ \text{H O}.$

338. $2\ (Pb\ O,\ [C_4\ H_3]\ O_3) + 2\ H\ O,\ C_8\ H_4\ O_{10} + Aq =$

$2\ \text{Pb O},\ \overset{\text{White.}}{\text{C}_8}\ \text{H}_4\ O_{10} + 2\ (H\ O,\ [C_4\ H_3]\ O_3) + Aq.$

$Pb\ O,\ N\ O_5 + [N\ H_4]\ O,\ H\ O + Aq = \overset{\text{White.}}{\text{Pb}}\ \text{O},\ \text{H O}$

$+ [N\ H_4]\ O,\ N\ O_5 + Aq.$

339. $Pb\ O,\ [C_4\ H_3]\ O_3 + Aq + 2\ \text{Pb O} =$

$3\ Pb\ O,\ [C_4\ H_3]\ O_3 + Aq.$

$3\ Pb\ O,\ [C_4\ H_3]\ O_3 + Aq + 2\ \mathbf{C\ O_2} =$

$2\ (\text{Pb}\ \overset{\text{White.}}{\text{O}},\ \text{C}\ O_2) + Pb\ O,\ [C_4\ H_3]\ O_3 + Aq.$

340. $\text{Zn} + Pb\ O,\ [C_4\ H_3]\ O_3 + Aq = \text{Pb}$

$+ Zn\ O,\ [C_4\ H_3]\ O_3 + Aq.$

341. $Pb\ O,\ [C_4\ H_3]\ O_3 + H\ S + Aq = \overset{\text{Black.}}{\text{Pb}}\ \text{S}$

$+ H\ O,\ [C_4\ H_3]\ O_3 + Aq.$

342. $\text{Pb S} + 3\ \mathbf{O} = \text{Pb O} + \mathbf{S\ O_2},$ also Pb S

$+ 4\ \mathbf{O} = \text{Pb}\ \overset{\text{White.}}{\text{O}},\ \text{S}\ O_3.$

342. By roasting galena we obtain a mixture of Pb O with a small amount of Pb O, S O_3. By then melting together Pb O or Pb O, S O_3 and an excess of Pb S, we obtain metallic lead and sulphurous acid, thus : —

$$2\,Pb\,O + Pb\,S = 3\,Pb + \mathbf{S\,O_2}, \text{ also } Pb\,O, S\,O_3 + Pb\,S = 2\,Pb + 2\,\mathbf{S\,O_2}.$$

Bismuth (Bi).

347. $$2\,Bi + 4\,(H\,O, \; N\,O_5) + aq = Bi_2\,O_3, 3\,N\,O_5 + aq + \mathbf{N\,O_2}.$$

$$4\,(Bi_2\,O_3, \; 3\,N\,O_5) + Aq = 3\,(\overset{\text{White.}}{Bi_2\,O_3}, \; N\,O_5) + Bi_2\,O_3, 9\,N\,O_5 + Aq.$$

$$Bi_2\,O_3, 9\,N\,O_5 + 3\,H\,S + Aq = \overset{\text{Brownish-Black.}}{Bi_2\,S_3} + 9\,(H\,O, \; N\,O_5) + Aq.$$

Copper (Cu).

349. Malachite $= \qquad Cu\,O, \overset{\text{Green.}}{H\,O.}\; Cu\,O, C\,O_2.$

Blue Carbonate of Copper $=$

$$Cu\,O, H\,O. \overset{\text{Blue.}}{2}\; Cu\,O, C\,O_2.$$

350. The oxides of copper are $\overset{\text{Red.}}{Cu_2\,O}$ and $\overset{\text{Black.}}{Cu\,O}.$

352. $$\overset{\text{Blue Solution.}}{Cu\,O, S\,O_3} + K\,O, H\,O + Aq = Cu\,\overset{\text{Blue.}}{O}, H\,O + K\,O, S\,O_3 + Aq.$$

352.   By boiling, $\overset{\text{Blue.}}{Cu\ O},\ H\ O + Aq = \overset{\text{Black.}}{Cu\ O} + Aq.$

353. $\overset{\text{Light Blue Solution.}}{Cu\ O,\ S\ O_3} + 2\,([N\,H_4]\ O,\ H\ O) + Aq =$

$\overset{\text{Very deep Blue Solution.}}{Cu\ O,\ S\ O_3.\ 2\ N\ H_3.\ H\ O} + Aq =$

$\left\{ N\ \overset{H_3}{Cu} \right\}\ O,\ S\ O_3.\ [N\,H_4]\ O,\ H\ O + Aq,$ a

solution which, when treated with alcohol, yields

crystals having the composition

$\left\{ N\ \overset{H_3}{Cu} \right\}\overset{\text{Deep Blue.}}{O,\ S\ O_3.\ [N\ H_4]\ O,\ H\ O.}$

These, when heated, are resolved into

$\left\{ N\ \overset{H_3}{Cu} \right\}\overset{\text{Green Powder.}}{O,\ S\ O_3} + \mathbf{[N\ H_4]\ O,\ H\ O.}$

354. $2\,(Cu\ O,\ S\ O_3) + 2\,(K\ O,\ H\ O) + Aq - \mathbf{O}^* =$
$\overset{\text{Yellow-Red.}}{Cu_2\ O} + 2\,(K\ O,\ S\ O_3) + Aq.$

355. $2\,(Cu\ O,\ S\ O_3) \overset{\text{Heated together.}}{+\ 2\,(Na\ O,\ C\ O_2)} + C = 2\ Cu$
$+ 2\,(Na\ O,\ S\ O_3) + 3\ \mathbf{C\ O_2}.$

356. $Zn + Cu\ O,\ S\ O_3 + Aq = Cu + Zn\ O,\ S\ O_3$
$+ Aq.$

357. $Cu\ O,\ H\ O + \mathbf{H} = Cu + 2\ \mathbf{H\ O}.$

359. $\overset{\text{Black.}}{Cu\ O} + H\ Cl + Aq = \overset{\text{Green Solution.}}{Cu\ Cl} + Aq.$

360. $3\ Cu + 4\,(H\ O,\ N\ O_5) + aq = 3\,(\overset{\text{Blue Solution.}}{Cu\ O,\ N\ O_5})$
$+ aq + \mathbf{N\ O_2}.$

* The oxygen is removed by adding grape sugar to the solution.

360. $2 \text{(Cu O, } \overset{\text{Blue Salt.}}{\text{N O}_5}.\ 4\text{ H O)} + \text{Sn} = 2 \overset{\text{Black.}}{\text{Cu O}} + \overset{\text{White.}}{\text{Sn O}_2}$

$+ 8\ \mathbf{H\ O} + 2\ \mathbf{N\ O_4}.$

$2\ (Cu\ O,\ S\ O_3) + 2\ (Na.O,\ C\ O_2) + Aq =$

$\text{Cu O, H } \overset{\text{Blue or Green.}}{\text{O. Cu O, C O}_2} + 2\ (Na\ O,\ S\ O_3)$

$+ Aq + \mathbf{C\ O_2}.$

$3\ (Cu\ O,\ S\ O_3) + [\dot{H}\ O,\ 2\ Na\ \dot{O}],\ P\ O_5 + Aq =$

$\overset{\text{Greenish-Blue.}}{3\text{ Cu O, P O}_5} + 2\ (Na\ O,\ S\ O_3) + H\ O,\ S\ O_3$

$+ Aq.$

Dioptase $= 3 \text{ Cu O,} \overset{\text{Emerald-Green.}}{2 \text{ Si O}_3}.\ 3\text{ H O}.$

361. Basic Acetate of Copper $=$

$\text{Cu O, } [\text{C}_4\text{ H}_3] \overset{\text{Green.}}{\text{O}_3}.\text{ Cu O, H O. 5 H O}.$

Neutral Acetate of Copper $=$

$\text{Cu O } [\overset{\text{Green.}}{\text{C}_4\text{ H}_3}] \text{O}_3.\ 5\text{ H O}.$

362. $\overset{\text{Blue Solution.}}{Cu\ O,\ S\ O_3} + H\ S + Aq = \overset{\text{Black.}}{\text{Cu S}} + H\ O,\ S\ O_3$

$+ Aq.$

$\overset{\text{Black.}}{\text{Cu S}} + H\ Cl + aq = \overset{\text{Green Solution.}}{Cu\ Cl} + aq + \mathbf{H\ S}.$

Mercury (Hg).

367. $6\ Hg + 4\ (H\ O,\ N\ O_5) + aq = 3\ (Hg_2\ O,\ N\ O_5)$

$+ aq + \mathbf{N\ O_2}.$

368. $Hg_2\ O,\ N\ O_5 + K\ O,\ H\ O + Aq = \overset{\text{Black.}}{Hg_2\ O}$

$+ K\ O,\ N\ O_5 + Aq.$

368. Hahnemann's Suboxide of Mercury has not a constant composition. Some chemists, however, assign to it the symbol $2\ Hg_2\ O,\ N\ O_5.\ N\ H_3$.

370. $Hg_2\ O,\ N\ O_5 + Na\ Cl + Aq = \overset{\text{White.}}{Hg_2}\ Cl + Na\ O,\ N\ O_5 + Aq.$

371. $3\ Hg + 4\ (H\ O,\ N\ O_5) + aq = 3\ (Hg\ O,\ N\ O_5) + aq + \mathbf{N\ O_2}.$

$Hg\ O,\ N\ O_5 + K\ O,\ H\ O + Aq = \overset{\text{Yellow.}}{Hg}\ O + K\ O,\ N\ O_5 + Aq.$

372. $\overset{\text{When heated.}}{Hg\ O,}\ N\ O_5 = \overset{\text{Red.}}{Hg}\ O + \mathbf{N\ O_4} + \mathbf{O}.$

373. $Hg\ O,\ N\ O_5 + Na\ Cl + Aq$, gives no precipitate, because Hg Cl is soluble in water.

$Hg\ O + H\ Cl + aq = \overset{\text{White.}}{Hg}\ Cl + \overset{\cdot}{H}\ O + aq.$

$Hg\ O, \overset{\text{Heated together.}}{S\ O_3} + Na\ Cl = \overset{\text{Residue.}}{Na}\ O,\ S\ O_3 + \overset{\text{Sublimate.}}{Hg}\ Cl.$

$Hg_2\ Cl + K\ O,\ H\ O + Aq = \overset{\text{Black.}}{Hg_2}\ O + K\ Cl + Aq.$

$Hg\ Cl + K\ O,\ H\ O + Aq = \overset{\text{Yellow.}}{Hg}\ O + K\ Cl + Aq.$

374. $2\ Hg\ Cl + [N\ H_4]\ O,\ H\ O + Aq = \left\{\overset{\text{White.}}{N}\genfrac{}{}{0pt}{}{H_2}{Hg_2}\right\}\ Cl + H\ Cl + Aq.$

375. $Hg\ Cl + Sn\ Cl + Aq = \overset{\text{Black Powder.}}{Hg} + Sn\ Cl_2 + Aq.$

376. $Hg\ Cl + H\ S + Aq = \overset{\text{Black.}}{Hg}\ S + H\ Cl + Aq.$

Silver (Ag).

380. $3\,Ag + 4\,(HO, N\,O_5) + Aq = 3\,(Ag\,O, N\,O_5)$
$+ Aq + \mathbf{N\,O_2}.$

381. *a.* $\overset{\text{Heated together.}}{Ag\,O, N\,O_5 + 2\,C} = Ag + 2\,\mathbf{C\,O_2} + \mathbf{N\,O_2}.$

     b. $Ag\,O, N\,O_5 + K\,O, H\,O + Aq = \overset{\text{Brown.}}{Ag\,O,}\,H\,O$
$+ K\,O, N\,O_5 + Aq.$

      $Ag\,O,\,H\,O + [N\,H_4]\,O, H\,O + Aq = \left\{ N \overset{\text{White.}}{\genfrac{}{}{0pt}{}{H_2}{Ag}} \right\}$
$+ Aq.$

     c. $Ag\,O, \ N\,O_5 \ + \ Na\,Cl \ + \ Aq \ = \ \overset{\text{White.}}{Ag}\,Cl$
$+ Na\,O, N\,O_5 + Aq.$

     e. $Ag\,O, N\,O_5 + H\,S + Aq = \overset{\text{Black.}}{Ag}\,S + H\,O, N\,O_5$
$+ Aq.$

Gold (Au).

385. $Au + 3\,H\,Cl + H\,O, N\,O_5 + Aq = Au\,Cl_3$
$+ Aq + \mathbf{N\,O_2}.$

387. $Au\,Cl_3 + 6\,(Fe\,O, S\,O_3) + Aq = \overset{\text{Dark-Brown.}}{Au} + Fe_2\,Cl_3$
$+ 2\,(Fe_2\,O_3, 3\,S\,O_3) + Aq.$

388.   Aurate of Potassa $(K\,O, \mathbf{Au\,O_3} + 4\,H\,O)$ is a
compound of oxide of potassium and teroxide
of gold, in which the last plays the part of an
acid.

Platinum (Pt).

391. $$3 \text{ Pt} + 6 \, H \, Cl + 2 \, (H \, O, \, N \, O_5) + Aq =$$
$$3 \, Pt \, Cl_2 + Aq + 2 \, \mathbf{N} \, \mathbf{O}_2.$$

392. $$[N \, H_4] \, Cl + Pt \, Cl_2 + Aq = [\text{N} \, \text{H}_4] \overset{\text{Yellow.}}{\text{Cl}}, \, \text{Pt} \, \text{Cl}_2$$
$$+ \, Aq.$$

394. $$K \, Cl + Pt \, Cl_2 + Aq = \text{K} \, \overset{\text{Yellow.}}{\text{Cl}}, \, \text{Pt} \, \text{Cl}_2 + Aq.$$

The Oxides of Platinum are $\text{Pt} \, \text{O}^*$ and $\overset{\text{Black.}}{\text{Pt}} \, \text{O}_2.$

The Chlorides of Platinum are $\overset{\text{Greenish-Brown.}}{\text{Pt} \, \text{Cl}}$ and $\overset{\text{Reddish-Brown.}}{\text{Pt} \, \text{Cl}_2}.$

The Sulphides of Platinum are $\overset{\text{Black.}}{\text{Pt} \, \text{S}}$ and $\overset{\text{Black.}}{\text{Pt} \, \text{S}_2}.$

$$Pt \, Cl + H \, S + Aq = \overset{\text{Black.}}{\text{Pt} \, \text{S}} + H \, Cl + Aq.$$

$$Pt \, Cl_2 + 2 \, H \, S + Aq = \overset{\text{Black.}}{\text{Pt} \, \text{S}_2} + 2 \, H \, Cl + Aq.$$

THIRD GROUP OF THE HEAVY METALS.

Chromium (Cr).

397. The symbol of chrome iron ore is $\text{Fe} \, \text{O}, \, \text{Cr}_2 \, \text{O}_3$; but almost invariably a portion of the $\text{Fe} \, \text{O}$ is replaced by $\text{Mg} \, \text{O}$, and a portion of the $\text{Cr}_2 \, \text{O}_3$ by $\text{Al}_2 \, \text{O}_3.$

$$2 \, (\text{Fe} \, \text{O}, \, \overset{\text{Heated together in contact with air.}}{\text{Cr}_2 \, \text{O}_3}) + 2 \, (\text{K} \, \text{O}, \, \text{C} \, \text{O}_2) + 7 \, \mathbf{O} =$$
$$\text{Fe}_2 \, \text{O}_3 + 2 \, (\text{K} \, \text{O}, \, 2 \, \text{Cr} \, \text{O}_3) + 2 \, \mathbf{C} \, \mathbf{O}_2.$$

* Only known in combination with water.

10

397.   The above process is hastened by mixing with the pulverized mineral a portion of nitre, which yields when heated a large supply of oxygen.

398. $KO, 2 \overset{\text{Red.}}{Cr} O_3 + KO, CO_2 + Aq = 2 (K\overset{\text{Yellow.}}{O, Cr} O_3)$
$+ Aq + \mathbf{C\,O_2}.$

$2 (K\overset{\text{Yellow.}}{O, Cr} O_3) + HO, NO_5 + Aq = K\overset{\text{Red.}}{O, 2} Cr O_3$
$+ KO, NO_5 + Aq.$

399. $KO, Cr O_3 + Pb O, [C_4 H_3] O_3 + Aq =$
$\overset{\text{Yellow.}}{Pb O, Cr O_3} + KO, [C_4 H_3] O_3 + Aq.$

$2 (Pb \overset{\text{Yellow.}}{O, Cr} O_3) + KO, HO + Aq = 2 Pb \overset{\text{Red.}}{O, Cr} O_3$
$+ KO, Cr O_3 + Aq.$

400. $2 (Pb \overset{\text{Yellow.}}{O, Cr} O_3) + 8 HCl + Aq = 2 \overset{\text{White.}}{Pb} Cl$
$+ \overset{\text{Green.}}{Cr_2 Cl_3} + Aq + 3 \mathbf{Cl}.$

$Cr_2 Cl_3 + 3 ([N H_4] O, HO) + Aq =$
$Cr_2 O_3, 10 HO + 3 [N H_4] Cl + Aq.$

$KO, 2 Cr O_3 + HO, S O_3 + 3 S O_2 + Aq =$
$KO, S O_3. Cr_2 O_3, 3 S O_3 + Aq.$

The symbol of crystallized chrome alum is

$KO, S O_3. Cr_2 O_3, 3 S O_3. 24 HO.$

401. $KO, 2 Cr O_3 + x* (HO, S O_3) + aq = 2 \overset{\text{Red.}}{Cr} O_3$
$+ KO, S O_3. HO, S O_3 + x (HO, S O_3)$
$+ aq.$

---

* x is here used to express an indefinite amount.

401. *a.* $2 \overset{\text{Red.}}{Cr} O_3 - 3\ \mathbf{O} = \overset{\text{Green.}}{Cr_2} O_3.$

The oxygen in the last reaction may be removed by alcohol or any other reducing agent.

Antimony (Sb).

The oxides of antimony are as follows : —

403.   Oxide of Antimony, $\overset{\text{White.}}{Sb}\ O_3$.

Antimonious Acid, $\overset{\text{White.}}{Sb}\ O_4 = \tfrac{1}{2}\ (Sb\ O_3,\ Sb\ O_5)$.

Antimonic Acid, $\overset{\text{Pale-Yellow.}}{Sb}\ O_5$.

404. $3\,Sb + 4\,(HO,\,N O_5) = 3\,SbO_4 + 4\,\mathbf{H}\,\mathbf{O} + 4\,\mathbf{N}\,\mathbf{O_2}.$

When antimony is treated with an excess of concentrated nitric acid, only $Sb\ O_4$ appears to be formed. If, however, the acid is dilute, the antimonious acid is mixed with more or less of basic nitrate of antimony $(2\ Sb\ O_3,\ N\ O_5)$ according to the degree of dilution.

By heating together one part of metallic antimony and four parts of nitre in a crucible, there is formed a white mass, which is a mixture of antimoniate of potassa $(K\ O,\ Sb\ O_5)$; nitrite of potassa $(K\ O,\ N\ O_3)$; and undecomposed nitre $(K\ O,\ N\ O_5)$. Warm water will dissolve the two last, but not the antimoniate of potassa. If, however, this anhydrous salt is boiled with water for one or two hours, it combines with five equivalents of water, forming a soluble compound $(K\,O'\,Sb\,O_5.\,5\,H\,O)$. The white mass, which seemed at first insoluble, dissolves in great measure, leaving in suspension only a small amount of binantimoniate of potassa.

405. $$Sb\,S_3 + 3\,H\,Cl + aq = Sb\,Cl_3 + aq + 3\,\mathbf{H\,S}.$$

$$Sb + 3\,H\,Cl + (H\,O,\ N\,O_5) + aq = Sb\,Cl_3 + aq + \mathbf{N\,O_2}.$$

$$Sb + x\,\mathbf{Cl} = Sb\,Cl_5 + x\,\mathbf{Cl}.$$

$$Sb\,Cl_3 + Aq = Sb\,O_3 + 3\,H\,Cl + Aq.$$

The precipitate which is first formed on diluting a concentrated solution of Sb Cl_3 with water, always contains some chloride with the oxide; but by continued washing with water, or still better, with a weak solution of Na O, C O_2, the whole will be converted into oxide.

$$Sb\,Cl_5 + Aq = Sb\,O_5 + 5\,H\,Cl + Aq.$$

406. $$Sb\,O_3 + [H\,O,\ K\,O]\ C_8\,H_4\,O_{10} + Aq = [K\,O,\ Sb\,O_3]\ C_8\,H_4\,O_{10} + Aq.$$

The symbol of crystallized tartar emetic is

$$[K\,O,\ Sb\,O_3]\ C_8\,H_4\,O_{10}.\ 2\,H\,O.$$

407. $[K\,O,\ Sb\,O_3]\ C_8\,H_4\,O_{10} + 3\,H\,S + Aq = \overset{\text{Red.}}{\text{Sb}}\,S_3$

$[H\,O,\ K\,O]\ C_8\,H_4\,O_{10} + Aq.$

Kermes mineral is an amorphous modification of Sb S$_3$.

$Sb\,Cl_5 + 5\,H\,S + Aq = \overset{\text{Bright-Yellow.}}{\text{Sb}}\,S_5 + 5\,H\,Cl + Aq.$

Golden Sulphuret is a mixture of Sb S$_3$ and Sb S$_5$.

408. $\overset{\text{Melted together.}}{\text{Sb S}_3} + 3\,\text{Fe} = \text{Sb} + 3\,\text{Fe S}.$

Arsenic (As).

412. $\text{As} + 3\,\mathbf{O} = \text{As O}_3$ (arsenious acid).

413. $2\,\text{As O}_3 + 3\,\text{C} = 2\,\text{As} + 3\,\mathbf{C O_2}.$

414. $\text{As O}_3 + K\,O,\ H\,O + Aq = K\,O,\ \text{As O}_3 + Aq.$

a. $K\,O,\ As\,O_3 + 2\,(Cu\,O,\ S\,O_3) + Aq =$

$2\,\overset{\text{Green.}}{\text{Cu O}},\ \text{As O}_3 + K\,O,\ S\,O_3.\ H\,O,\ S\,O_3 + Aq.$

b. The symbol of Schweinfurth green is

$\text{Cu O},\ [C_4\,H_3]\,O_3.\ 2\,\text{Cu O},\ \text{As O}_3.$

415. $\text{As O}_3 + 2\,(H\,O,\ N\,O_5) + a\acute{q} = As\,O_5 + aq$

$+ 2\,\mathbf{N\,O_4}.$

The symbol of crystallized binarseniate of potassa is $[2\,\text{H O},\ \text{K O}]\,\text{As O}_5.$

416. $As\,O_3 + 3\,H\,S + Aq = \overset{\text{Yellow.}}{\text{As}}\,S_3 + Aq.$

$As\,O_5 + 5\,H\,S + Aq = \overset{\text{Yellow.}}{\text{As}}\,S_5 + Aq.$

$2\,\overset{\text{Yellow.}}{\text{As S}_3} + S = 2\,\overset{\text{Red.}}{\text{As}}\,S_2.$

10 *

416.   The symbol of Mispickel (arsenical pyrites) is
Fe [As, S_2].

$$2 \,(Fe\,[As,\; S_2]) + 13\,\mathbf{O} = Fe_2\,O_3 + 2\;\mathbf{As}\,\mathbf{O_3}$$
$$+\,2\;\mathbf{S}\;\mathbf{O_2}.$$

417. $As\,Zn_3 + 3\,(H\,O,\; S\,O_3) + Aq = 3\,(Zn\,O,\; S\,O_3)$
$+\,Aq + \mathbf{As}\;\mathbf{H_3}.$

418. $Sb\,Zn_3 + 3\,(H\,O,\; S\,O_3) + Aq = 3\,(Zn\,O,\; S\,O_3)$
$+\,Aq + \mathbf{Sb}\;\mathbf{H_3}.$

The compounds of arsenic and antimony with hydrogen are always mixed with more or less free
hydrogen.   When prepared as described in
sections 417, 418 of Stöckhardt's Elements, the
gas consists almost entirely of hydrogen, containing only a very minute amount of either metallic compound.

TABLES.

EXPLANATION OF TABLES.

TABLE I. — This table, which has been reprinted from the "Elementary Instructions in Chemical Analysis" by Fresenius, indicates by means of figures the solubility or insolubility in water and acids of some of the more frequently occurring compounds; thus, 1 means a substance soluble in water; 2, a substance insoluble in water, but soluble in chlorohydric or nitric acid; 3, a substance insoluble either in water or acids. For those substances standing on the limits between these three classes, the figures are jointly expressed; thus 1 - 2 signifies a substance difficultly soluble in water, but soluble in chlorohydric or nitric acid; 1 - 3, a body difficultly soluble in water, and the solubility of which is not increased on the addition of acids; and 2 - 3, a substance insoluble in water, and difficultly soluble in the acids. When the relation of a substance to hydrochloric acid is different from that to nitric acid, this is stated in the notes. The figure indicating the solubility of a given salt will be found opposite to the symbol of its acid, in the column headed by the symbol of its base; that of a given binary, under the symbol of the corresponding oxide, and opposite to the symbol of its electro-negative element.

TABLE II. — The values of the French measures and weights, in terms of the corresponding English units, given in this table, were taken from the second volume of the Cavendish Edition of "Gmelin's Hand-Book of Chemistry." The logarithms of these values and their arithmetical complements have been added to facilitate the reduction from one system to the other. The use of the table can be illustrated best by a few examples.

1. It is required to reduce 560.367 metres to English feet.

Solution. — No. of feet = No. of metres $\times$ No. of feet in one metre.

log. No. of feet = log. No. of metres + log. No. of feet in one metre.

log. 560.367	2.7484726
log. 3.2809 (Value in feet of one metre from Table II.)	0.5159930
	3.2644656

Ans. = 1838.51 feet.

2. It is required to reduce 30.964 inches to centimetres.

Solution. — No. of centimetres = No. of inches ÷ No. of inches in one cen-
        timetre.

    log. No. of centimetres = log. No. of inches + log. (ar. co.) No. of
        inches in one centimetre.

    log. 30.964   ·   1.4908571
    log. (ar. co.) 0.3937 (Value of one centimetre in inches)   0.4048258
                                                                     —————————
                                                                     1.8956829

            Ans. 78.6471 centimetres.

3. It is required to reduce 23.576 kilometres to feet.

Solution. — 23.576 kilometres = 23576 metres.

    No. of feet = No. of metres × No. of feet in one metre.
    log. 2.3576   4.3724701
    log. 3.2809   0.5159930
                                                                     —————————
                                                                     4.8884631

            Ans. 77350.5 feet.

In the above examples logarithms of seven places have been used; but where
great accuracy is not required, logarithms of four places are sufficient. In
such cases the last three figures of the logarithm given in the table may be
neglected, and the problems solved with great expedition by means of the table
of four-place logarithms which accompanies this book.

TABLE III. — This table has been taken, with some few alterations, from
Weber's "Atomgewichts-Tabellen." The atomic volumes assigned to the elements are the same as those generally given in English and American text-
books on Chemistry, with the exception of those of Carbon, Boron, and Silicon,
which are assumed to yield a one-volume gas like oxygen for convenience in
calculation. The calculated specific gravities are deduced from the observed
specific gravity of oxygen and the chemical equivalent of the given substance
by means of the proportion,

Equiv. of Oxygen : Equiv. of given substance = Sp. Gr. of Oxygen : Sp. Gr.
                    of given substance.

This proportion yields the specific gravity directly when one equivalent of
the substance occupies the same volume as one equivalent of oxygen. If it
occupies twice, three times, or four times this volume, the results must be
divided by two, three, or four, as the case may be. The method of calculating
may best be illustrated by a few examples.

1. It is required to calculate the Specific Gravity of Nitrogen.

                 Equiv. of O. Equiv. of N.   Sp. Gr. of O.
    Solution.   8   :   14   =   1.10563 :   1.93485.

This would be the specific gravity if 14 parts of nitrogen occupied the same
volume as 8 parts of oxygen; or, in other words, if the equivalent volume of

nitrogen was 1, the same as that of oxygen. The fact is that it is 2, so that the true specific gravity of nitrogen $= \frac{1}{2} (1.93485) = 0.967428$.

2. It is required to calculate the specific gravity of ammonia gas.

	Equiv. of O.	Equiv. of N H$_3$.	Sp. Gr. of O.	
Solution.	8 :	17 $=$	1.10563 :	2.34946.

Hence the specific gravity of ammonia gas would be 2.34946 if one equivalent (or 17 parts) occupied the same volume as one equivalent of oxygen; but on referring to the table, it will be found that the equivalent volume of this gas is 4, or, in other words, that one equivalent occupies a volume four times as large as that occupied by one equivalent of oxygen, so that to find the true specific gravity of ammonia gas we must divide 2.34946 by 4, which will give a quotient 0.58736. The slight difference between this result and that given in the table arises from the fact that in Weber's table the equivalent of nitrogen used is 14.005, and not 14, as in the solutions of the above examples.

Were the law of equivalent volumes absolutely rigorous, (that is, did one equivalent of every gas precisely occupy either the same volume, or else a volume two, three, or four times as great as the volume of oxygen,) then the calculated specific gravities ought to agree exactly with those obtained by experiment. On comparing together the two columns of observed and calculated specific gravities in the table, it will be found that the numbers, although approximatively equal, do not absolutely coincide. Part of these differences are unquestionably owing to errors of observation; but after making the greatest possible allowance for all errors of that sort, there still remains (especially in the case of the easily condensed gases, such as alcohol vapor, sulphurous acid, and carbonic acid) large differences to be accounted for. The most probable explanation of these differences seems to be found in the assumption that the law of equivalent volumes holds rigorously only when the gases are in the state of extreme expansion. As we experiment upon them, they are more or less condensed by the pressure of the atmosphere, and it is supposed that they are not all condensed equally, or, in other words, that even under this pressure they do not obey absolutely the law of Marriotte. The more easily a gas may be reduced to a fluid, the greater is it condensed by the atmospheric pressure, and hence the greater is its specific gravity. This view is confirmed by the fact that the observed specific gravity of carbonic acid gas at 0°, and under a more feeble pressure than that of the atmosphere, approaches more nearly to that obtained by experiment.

The Specific gravity of Carbonic Acid Gas, at 0° (air $= 1$), was,

Under the pressure of 76.000 centimetres,	1.52910
" " " 37.413 "	1.52366
" " " 22.417 * "	1.52145
Theoretical specific gravity,	1.52024

---

* It must be remembered that the specific gravity of a gas is equal to its weight, divided by the weight of an equal volume of air under the same conditions of temperature and pressure.

TABLE I.

SOLUBILITY OF FREQUENTLY OCCURRING COMPOUNDS.

	$[NH_4]O$	NaO	KO	AgO	CaO	SrO	BaO	PbO	MgO	ZnO	CdO	CuO	HgO	Hg_2O
Cl	1	1	1	1—2	1—2	1	1	2	2	2	2	2	2	2
I	1	1	1	3	1	1	1	1—3	1	1	1	1	1	2
SO_3	1	1	1	3	1—2	1	1	2	1	1	1	2†	2	2—3
CrO_3	1	1	1	2*	1—3	3	3	3	2	2	2		3	3
NO_5	1	1	1	1—3	1	2	2	2—3	1	1	1	1	1—2	—
PO_5	1	1	1	2	1	1	1	1	1			2		2
AsO_5	1	1	1	1	1	2	2	2	1	1	1	1	1	1
AsO_3	1	1	1	2	2	2	2	2	2	2	2	2	2	2
CO_2	1	1	1	2	2	1	2	2	2	3	2	2	2	2
BO_3	1	1	1	2	2	2	2	2	2	2	2	2	2	2
C_2O_3	1	1	1	2	2	2	2	2	2	2	1—2	2	2	1—2
$\overline{A}$	1	1	1	1	1	1	1	1	1	1	1	1	1	1—1
$\overline{T}$	1	1	1	2	2	2	2	2	1—2	2	1—2	1	2	2—1

	Mn O	Fe O	Ca O	Ni O	Al₂ O₃	Cr₂ O₃	Fe₂ O₃	Sn O	Sn O₂	Pt O₂	Au O₃	Sb O₃	Bi O₃
Cl	2	2	2	2	2 and 3‖	2 and 3‖	2 and 3‖	2	2 and 3‖	2		2†	2
I	1	1	1	1	1	1	1	1	1	1	1	1	1
S	1	1	2†	2†				2†	2†	3§		2†	2
S O₃	2	2	1	1	1	1	1	1	1	1		2	1
Cr O₃	1	1	2	1	2	2	1	2				2	2
N O₅	1	1	1	1	1	1	1			1			1
P O₅	1	2	2	1	2	2	1					2	
As O₅	2	2	2	2	2	1	2					2	2
As O₃	2	2	2	2	2	2	2	2				2	1
C O₂	2	2	2	2	2	2	2	2					2
B O₃	2	2	2	2	2	1	1—2	2				2	2
C₂ O₃	2	1—2	2					1				1—2	2
A		1	1	1	1	1	1	1	1			1	1
T	1—2	1—2	1				1	1—2				1	2

* Soluble in nitric acid, but not in hydrochloric.
† Soluble in nitric acid, but only very difficultly in hydrochloric.
‡ Decomposed and dissolved by hydrochloric, but converted into insoluble oxides by nitric acid.
§ Not affected by hydrochloric, but converted into a soluble sulphate by nitric acid.
‖ Soluble in one modification, and not in another.

TABLE II.

FRENCH MEASURES AND WEIGHTS.

MEASURES OF LENGTH.

1 Kilometre	=	1000 Metres.		1 Metre	=	1.000 Metre.
1 Hectometre	=	100 "		1 Decimetre	=	0.100 "
1 Decimetre	=	10 "		1 Centimetre	=	0.010 "
1 Metre	=	1 "		1 Millimetre	=	0.001 "

			Logarithms.	Ar. Co. Log.
1 Kilometre	=	0.6214 Miles.	9.7933712	0.2066188
1 Metre	=	3.2809 Feet.	0.5159930	9.4840070
1 Centimetre	=	0.3937 Inches.	9.5951742	0.4048258

MEASURES OF VOLUME.

1 Cubic Metre	=	1000.000 Litres.
1 Cubic Decimetre	=	1.000 "
1 Cubic Centimetre	=	0.001 "

			Logarithms.	Ar. Co. Log.
1 Cubic Metre	=	35.31660 Cubic Feet.	1.5479790	8.4520210
1 Cubic Decimetre	=	61.02709 Cubic Inches.	1.7855226	8.2144774
1 Cubic Centimetre	=	0.06103 " "	8.7855226	1.2144774
1 Litre	=	0.22017 Gallons.	9.3427581	0.6572419
1 Litre	=	0.88066 Quarts.	9.9448083	0.0551917
1 Litre	=	1.76133 Pints.	0.2458407	9.7541593

WEIGHTS.

1 Kilogramme	=	1000 Grammes.		1 Gramme	=	1.000 Gramme.
1 Hectogramme	=	100 "		1 Decigramme	=	0.100 "
1 Decagramme	=	10 "		1 Centigramme	=	0.010 "
1 Gramme	=	1 "		1 Milligramme	=	0.001 "

			Logarithms.	Ar. Co. Log.
1 Kilogramme	=	2.67951 Pounds (Troy).	0.4280554	9.5719446
1 Gramme	=	15.44242 Grains.	1.1887154	8.8112846

TABLE III.

SPECIFIC GRAVITY AND ABSOLUTE WEIGHT OF ONE
LITRE OF SOME OF THE MOST IMPORTANT GASES
AND VAPORS.

Names of Gases.	Equiv. Vol.	Sp. Gr. Obs'r'ed.	Sp. Gr. Calculated.	Weight of 1 Litre = 1000 c. c.	Logarithms.	Ar. Co. Logarithms.
Air,		1.	1.00000	1.29363	0.1118101	9.8881899
Alcohol,	4	1.613	1.58934	2.05602	0.3130273	9.6869727
Ammonia Gas,	4	0.5967	0.58753	0.76005	9.8808422	0.1191578
Antimony,	1		17.83274	23.06897	1.3630282	8.6369718
Antimonide of Hydr.	4		4.56239	5.90204	0.7710022	9.2289978
Arsenic,	1	10.65	10.36528	13.40884	1.1273912	8.8726088
Arsenide of Hydr.	4	2.695	2.69553	3.48702	0.5424519	9.4575481
Boron,	1		1.50591	1.94809	0.2896090	9.7103910
Bromine,	2	5.54	5.52605	7.14866	0.8542247	9.1457753
Bromohydric Acid,	4		2.79758	3.61903	0.5585922	9.4414078
Carbon,	1	0.8469*	0.82922	1.07270	0.0304783	9.9695217
Carbonic Acid,	2	1.52908	1.52024	1.96663	0.2937226	9.7062774
Carbonic Oxide,	2	0.96779	0.96743	1.25150	0.0974309	9.9025691
Chlorine,	2	2.47	2.45052	3.17007	0.5010689	9.4989311
Chloride of Boron,	4	3.942	4.05226	5.24213	0.7195078	9.2804922
Chloride of Silicon,	3	5.939	5.92477	7.66446	0.8844816	9.1155184
Chlorohydric Acid,	4	1.2474	1.25981	1.62973	0.2121157	9.7878843
Cyanogen,	2	1.8064	1.79698	2.32463	0.3663538	9.6336462
Cyanhydric Acid,	4	0.9476	0.93304	1.20701	0.0817109	9.9182891
Ether,	2	2.586	2.55677	3.30751	0.5195012	9.4804988
Fluorine,	2		1.30151	1.68367	0.2262570	9.7737430
Fluoride of Boron,	4	2.3124	2.32875	3.01254	0.4789329	9.5210671
Fluoride of Silicon,	3	3.600	3.62677	4.69170	0.6713302	9.3286698
Fluohydric Acid,	4		0.68531	0.88654	9.9476983	0.0523017
Hydrogen,	2	0 06927	0.06910	0.08939	8.9512889	1.0487111
Iodine,	2	8.716	8.76760	11.34203	1.0547011	8.9452989
Iodohydric Acid,	4	4.443	4.41835	5.71571	0.7570702	9.2429298
Marsh Gas,	4	0.5576	0.55282	0.71514	9.8543911	0.1456089
Mercury,	2	6.976	6.91732	8.94845	0.9517478	9.0482522
Nitrogen,	2	0.97136	0.96776	1.25192	0.0975765	9.9024235
Nitrous Oxide,	2	1.5269	1.58951	2.05624	0.3130738	9.6869262
Nitric Oxide,	4	1.0388	1.03669	1.34109	0.1274580	9.8725420
Olefiant Gas,	4	0.9852	0.96743	1.25150	0.0974309	9.9025691
Oxygen,	1	1.10563		1.43028	0.1554210	9.8445790
Phosphorus,	1	4.42	4.33452	5.60727	0.7487515	9.2512485
Phosphide of Hydr.	4	1.178	1.18728	1.53590	0.1863629	9.8136371
Selenium,	1		2.73801	3.54197	0.5492448	9.4507552
Silicon,	1		3.07120	3.97300	0.5991186	9.4008814
Sulphur,	⅓	6.5635	6.65866	8.61384	0.9351968	9.0648032
Sulphide of Hydr.	2	1.1912	1.17888	1.52503	0.1832784	9.8167216
Sulphurous Acid,	2	2.247	2.21541	2.86592	0.4572640	9.5427360

* Calculated from the Sp. Gr. of Carbonic Acid Gas, observed by Regnault.

TABLE IV.

MEAN COEFFICIENTS OF LINEAR EXPANSION OF SOLIDS FOR ONE DEGREE BETWEEN 0° AND 100°.

Name of Substance.	Coefficients.	Name of Observer.
Glass (flint of Choisy le Roi),	0.00000760	Regnault.
Platinum,	0.00000884	Dulong and Petit.
Glass (common of Paris),	0.00000920	Regnault.
Palladium,	0.00001000	Wollaston.
Antimony,	0.00001083	Smeaton.
Iron (soft forged),	0.00001220	Lavoisier and Laplace.
Bismuth,	0.00001392	Smeaton.
Gold,	0.00001466	Lavoisier and Laplace.
Brass (English, in rods),	0.00001893	Roy.
Copper,	0.00001919	Troughton.
Silver,	0.00002083	"
Tin (fine),	0.00002283	"
Lead,	0.00002866	"
Zinc,	0.00002942	"

APPARENT CUBIC EXPANSION OF LIQUIDS IN GLASS BETWEEN 0° AND 100°.

Name of Substance.	Expansion from 0° to 100°.		
Water,	$\frac{1}{22}$	=	0.0466
Chlorohydric Acid, Sp. Gr. 1.137,	$\frac{1}{27}$	=	0.0600
Nitric Acid, Sp. Gr. 1.40,	$\frac{1}{9}$	=	0.0100
Sulphuric Acid, Sp. Gr. 1.85,	$\frac{1}{17}$	=	0.0600
Common Ether,	$\frac{1}{14}$	=	0.0700
Olive Oil,	$\frac{1}{12}$	=	0.0800
Oil of Turpentine,	$\frac{1}{14}$	=	0.0700
Water saturated with salt,	$\frac{1}{20}$	=	0.0500
Alcohol,	$\frac{1}{9}$	=	0.1100
Mercury,	$\frac{1}{64}$	=	0.0156

COEFFICIENTS OF CUBIC EXPANSION OF GASES.
OBSERVED BY V. REGNAULT.

Name of Substance.	Under constant Volume.	Under constant Pressure.
Air,	0.003665	0.003670
Nitrogen,	0.003668	0.003670
Hydrogen,	0.003667	0.003661
Oxide of Carbon,	0.003667	0.003669
Carbonic Acid,	0.003688	0.003710
Cyanogen,	0.003829	0.003877
Sulphurous Acid,	0.003845	0.003903

TABLE V.

DENSITIES AND VOLUMES OF WATER.

By M. DESPRETZ.

Temperature.	Volumes.	Densities.	Temperature.	Volumes.	Densities.	Temperature.	Volumes.	Densities.
0	1.0001269	0.999873	34	1.00555	0.994480	68	1.02144	0.979010
1	1.0000730	0.999927	35	1.00593	0.994104	69	1.02200	0.978473
2	1.0000331	0.999966	36	1.00624	0.993799	70	1.02255	0.977947
3	1.0000083	0.999999	37	1.00661	0.993433	71	1.02315	0.977373
4	1.0000000	1.000000	38	1.00699	0.993058	72	1.02375	0.976800
5	1.0000082	0.999999	39	1.00734	0.992713	73	1.02440	0.976181
6	1.0000309	0.999969	40	1.00773	0.992329	74	1.02499	0.975619
7	1.0000708	0.999929	41	1.00812	0.991945	75	1.02562	0.975018
8	1.0001216	0.999878	42	1.00853	0.991542	76	1.02631	0.974364
9	1.0001879	0.999812	43	1.00894	0.991139	77	1.02694	0.973766
10	1.0002684	0.999731	44	1.00938	0.990707	78	1.02761	0.973132
11	1.0003598	0.999640	45	1.00985	0.990246	79	1.02823	0.972545
12	1.0004724	0.999527	46	1.01020	0.989903	80	1.02885	0.971959
13	1.0005862	0.999414	47	1.01067	0.989442	81	1.02954	0.971307
14	1.0007146	0.999285	48	1.01109	0.989032	82	1.03022	0.970666
15	1.0008751	0.999125	49	1.01157	0.988562	83	1.03090	0.970027
16	1.0010215	0.998979	50	1.01205	0.988093	84	1.03156	0.969405
17	1.0012067	0.998794	51	1.01248	0.987674	85	1.03225	0.968757
18	1.00139	0.998612	52	1.01297	0.987196	86	1.03293	0.968120
19	1.00158	0.998422	53	1.01345	0.986728	87	1.03351	0.967482
20	1.00179	0.998213	54	1.01395	0.986243	88	1.03430	0.966837
21	1.00200	0.998004	55	1.01445	0.985756	89	1.03500	0.966183
22	1.00222	0.997784	56	1.01495	0.985270	90	1.03566	0.965567
23	1.00244	0.997566	57	1.01547	0.984766	91	1.03639	0.964887
24	1.00271	0.997297	58	1.01597	0.984281	92	1.03710	0.964227
25	1.00293	0.997078	59	1.01647	0.983798	93	1.03782	0.963558
26	1.00321	0.996800	60	1.01698	0.983303	94	1.03852	0.962908
27	1.00345	0.996562	61	1.01752	0.982782	95	1.03925	0.962232
28	1.00374	0.996274	62	1.01809	0.982231	96	1.03999	0.961547
29	1.00403	0.995986	63	1.01862	0.981720	97	1.04077	0.960827
30	1.00433	0.995688	64	1.01913	0.981229	98	1.04153	0.960125
31	1.00463	0.995391	65	1.01967	0.980709	99	1.04228	0.959434
32	1.00494	0.995084	66	1.02025	0.980152	100	1.04315	0.958634
33	1.00525	0.994777	67	1.02085	0.979576			

TABLE VI.

PER CENT OF N O5 IN AQUEOUS SOLUTIONS OF DIFFERENT SPECIFIC GRAVITIES. By URE. 15° C.

Specific Gravity.	Per Cent of NO_5	Specific Gravity.	Per Cent of NO_5.	Specific Gravity.	Per Cent of NO_5.	Specific Gravity.	Per Cent of NO_5.
1.500	79.7	1.419	59.8	1.295	39.8	1.140	19.9
1.498	78.9	1.415	59.0	1.289	39.0	1.134	19.1
1.496	78.1	1.411	58.2	1.283	38.3	1.129	18.3
1.494	77.3	1.406	57.4	1.276	37.5	1.123	17.5
1.491	76.5	1.402	56.6	1.270	36.7	1.117	16.7
1.488	75.7	1.398	55.8	1.264	35.9	1.111	15.9
1.485	74.9	1.394	55.0	1.258	35.1	1.105	15.1
1.482	74.1	1.388	54.2	1.252	34.3	1.099	14.3
1.479	73.3	1.383	53.4	1.246	33.5	1.093	13.5
1.476	72.5	1.378	52.6	1.240	32.7	1.088	12.7
1.473	71.7	1.373	51.8	1.234	31.9	1.082	11.9
1.470	70.9	1.368	51.1	1.228	31.1	1.076	11.2
1.467	70.1	1.363	50.2	1.221	30.3	1.071	10.4
1.464	69.3	1.358	49.4	1.215	29.5	1.065	9.6
1.460	68.5	1.353	48.6	1.208	28.7	1.059	8.8
1.457	67.7	1.348	47.9	1.202	27.9	1.054	8.0
1.453	66.9	1.343	47.0	1.196	27.1	1.048	7.2
1.450	66.1	1.338	46.2	1.189	26.3	1.043	6.4
1.446	65.3	1.332	45.4	1.183	25.5	1.037	5.6
1.442	64.5	1.327	44.6	1.177	24.7	1.032	4.8
1.439	63.8	1.322	43.8	1.171	23.9	1.027	4.0
1.435	63.0	1.316	43.0	1.165	23.1	1.021	3.2
1.431	62.2	1.311	42.2	1.159	22.3	1.016	2.4
1.427	61.4	1.306	41.4	1.153	21.5	1.011	1.6
1.423	60.6	1.300	40.4	1.146	20.7	1.005	0.8

PER CENT OF H Cl IN AQUEOUS SOLUTIONS OF DIFFERENT SPECIFIC GRAVITIES. By E DAVY 15° C.

Specific Gravity.	Per Cent of H Cl.	Specific Gravity.	Per Cent of H Cl.
1.21	42.43	1 10	20.20
1.20	40.80	1.09	18.18
1.19	38.38	1.08	16.16
1.18	36.36	1.07	14.14
1.17	34.34	1.06	12.12
1.16	32.32	1.05	10.10
1.15	30.30	1.04	8.08
1.14	28.28	1.03	6.06
1.13	26.26	1.02	4.04
1.12	24.24	1.01	2.02
1.11	22.22		

PER CENT OF H O, S O₃ AND S O₃ IN AQUEOUS SOLUTIONS OF DIFFERENT SPECIFIC GRAVITIES. BY BINEAU. 15° C.

Per Cent of H O, S O₃.	Specific Gravity.	Per Cent of S O₃.	Per Cent of H O, S O₃	Specific Gravity.	Per Cent of S O₃.
100	1.8426	81.63	56	1.4586	45.71
99	1.842	80.81	55	1.448	44.89
98	1.8406	80.00	54	1.438	44.07
97	1.840	79.18	53	1.428	43.26
96	1.8384	78.36	52	1.418	42.45
95	1.8376	77.55	51	1.408	41.63
94	1.8356	76.73	50	1.398	40.81
93	1.834	75.91	49	1.3886	40.00
92	1.831	75.10	48	1.379	39.18
91	1.827	74.28	47	1.370	38.36
90	1.822	73.47	46	1.361	37.55
89	1.816	72.65	45	1.351	36.73
88	1.809	71.83	44	1.342	35.82
87	1.802	71.02	43	1.333	35.10
86	1.794	70.10	42	1.324	34.28
85	1.786	69.38	41	1.315	33.47
84	1.777	68.57	40	1.306	32.65
83	1.767	67.75	39	1.2976	31.83
82	1.756	66.94	38	1.289	31.02
81	1.745	66.12	37	1.281	30.20
80	1.734	65.30	36	1.272	29.38
79	1.722	64.48	35	1.264	28.57
78	1.710	63.67	34	1.256	27.75
77	1.698	62.85	33	1.2476	26.94
76	1.686	62.04	32	1.239	26.12
75	1.675	61.22	31	1.231	25.30
74	1.663	60.40	30	1.223	25.49
73	1.651	59.59	29	1.215	23.67
72	1.639	58.77	28	1.2066	22.85
71	1.637	57.95	27	1.198	22.03
70	1.615	57.14	26	1.190	21.22
69	1.604	56.32	25	1.182	20.40
68	1.592	55.59	24	1.174	19.58
67	1.580	54.69	22	1.159	17.95
66	1.578	53.87	20	1.144	16.32
65	1.557	53.05	18	1.129	14.69
64	1.545	52.24	16	1.1136	13.06
63	1.534	51.42	14	1.098	11.42
62	1.523	50.61	12	1.083	9.79
61	1.512	49.79	10	1.068	8.16
60	1.501	48.98	8	1.0536	6.53
59	1.490	48.16	6	1.039	4.89
58	1.480	47.34	4	1.0256	3.26
57	1.469	46.53	2	1.013	1.63

PER CENT OF N H₃ IN AQUEOUS SOLUTIONS OF DIFFERENT SPECIFIC GRAVITIES. By J. OTTO. 16° C.

Specific Gravity.	Per Cent of N H₃.	Specific Gravity.	Per Cent of N H₃.	Specific Gravity.	Per Cent of N H₃.
0.9517	12.000	0.9607	9.625	0.9697	7.250
0.9521	11.875	0.9612	9.500	0.9702	7.125
0.9526	11.750	0.9616	9.375	0.9707	7.000
0.9531	11.625	0.9621	9.250	0.9711	6.875
0.9536	11.500	0.9626	9.125	0.9716	6.750
0.9540	11.375	0.9631	9.000	0.9721	6.625
0.9545	11.250	0.9636	8.875	0.9726	6.500
0.9550	11.125	0.9641	8.750	0.9730	6.375
0.9555	11.000	0.9645	8.625	0.9735	6.250
0.9556	10.950	0.9650	8.500	0.9740	6.125
0.9559	10.875	0.9654	8.375	0.9745	6.000
0.9564	10.750	0.9659	8.250	0.9749	5.875
0.9569	10.625	0.9664	8.125	0.9754	5.750
0.9574	10.500	0.9669	8.000	0.9759	5.625
0.9578	10.375	0.9673	7.875	0.9764	5.500
0.9583	10.250	0.9678	7.750	0.9768	5.375
0.9588	10.125	0.9683	7.625	0.9773	5.250
0.9593	10.000	0.9688	7.500	0.9778	5.125
0.9597	9.875	0.9692	7.375	0.9783	5.000
0.9602	9.750				

TABLE VII.

PER CENT OF OXYGEN IN DIFFERENT OXIDES.

Oxides.	Per Cent.	Logarithms.	Oxides.	Per Cent.	Logarithms.
$Al_2 O_3$	0.467	9.6684	$K O$	0.170	9.2304
$As O_5$	0.348	9.5416	$Li O$	0.550	9.7404
$B O_3$	0.688	9.8376	$M O_3$	0.343	9.5353
$Ba O$	0.104	9.0170	$Mg O$	0.400	9.6021
$Be_2 O_3$	0.630	9.7993	$Mn O$	0.225	9.3522
$Bi_2 O_3$	0.103	9.0128	$Mn_2 O_3$	0.303	9.4814
$C O_2$	0.727	9.8615	$N O_5$	0.740	9.8692
$Ca O$	0.286	9.4564	$Na O$	0.258	9.4116
$Co O$	0.213	9.3284	$Ni O$	0.213	9.3284
$Cr_2 O_3$	0.473	9.6749	$P O_5$	0.563	9.7505
$Cu_2 O$	0.112	9.0492	$Pb O$	0.717	9.8555
$Cu O$	0.201	9.3032	$Si O_3$	0.530	9.7243
$Fe O$	0.222	9.3464	$Sr O$	0.154	9.1875
$Fe_2 O_3$	0.300	9.4771	$Ta O_3$	0.115	9.0607
$H O$	0.889	9.9489	$Zn O$	0.197	9.2945